The Ocean of Today, the Legacy of Tomorrow

Albert Calbet

The Ocean of Today, the Legacy of Tomorrow

Navigating the Future of Marine Life and Ecosystems

Albert Calbet
Institute of Marine Sciences - CSIC
Barcelona, Spain

ISBN 978-3-031-88270-8   ISBN 978-3-031-88271-5   (eBook)
https://doi.org/10.1007/978-3-031-88271-5

© The Editor(s) (if applicable) and The Author(s), under exclusive license to Springer Nature Switzerland AG 2025

This work is subject to copyright. All rights are solely and exclusively licensed by the Publisher, whether the whole or part of the material is concerned, specifically the rights of translation, reprinting, reuse of illustrations, recitation, broadcasting, reproduction on microfilms or in any other physical way, and transmission or information storage and retrieval, electronic adaptation, computer software, or by similar or dissimilar methodology now known or hereafter developed.
The use of general descriptive names, registered names, trademarks, service marks, etc. in this publication does not imply, even in the absence of a specific statement, that such names are exempt from the relevant protective laws and regulations and therefore free for general use.
The publisher, the authors and the editors are safe to assume that the advice and information in this book are believed to be true and accurate at the date of publication. Neither the publisher nor the authors or the editors give a warranty, expressed or implied, with respect to the material contained herein or for any errors or omissions that may have been made. The publisher remains neutral with regard to jurisdictional claims in published maps and institutional affiliations.

This Springer imprint is published by the registered company Springer Nature Switzerland AG
The registered company address is: Gewerbestrasse 11, 6330 Cham, Switzerland

If disposing of this product, please recycle the paper.

Introduction

The Changing Ocean

The ocean, vast and seemingly infinite, has long been a symbol of mystery, power, and constancy. For millennia, it has acted as Earth's regulator, absorbing heat, storing carbon, and providing sustenance for countless species, including humans. Its rhythms, dictated by tides and currents, have shaped civilizations, economies, and ecosystems. Yet, beneath the surface of this familiar expanse, the ocean is undergoing profound transformations that will define the future of our planet.

The ocean of tomorrow will not be the ocean of today, nor of yesterday. Environmental forces, both natural and anthropogenic, are pushing marine ecosystems toward unprecedented states of change. These transformations will influence everything from microscopic plankton drifting in the water column to the great whales migrating across continents. The very fabric of the ocean's intricate web of life is being altered at a scale and pace that challenges our understanding. To comprehend these changes is to prepare for a world where the boundaries of the known ocean expand into new, uncertain realities.

Human activity has been one of the most significant drivers of these changes in the past 100 years. For centuries, we have viewed the ocean as both an inexhaustible resource and an indestructible force. Fishing fleets scoured its depths, believing its abundance to be limitless. Industrial activity poured pollutants into its waters, assuming the ocean's vastness would dilute and cleanse. The extraction of oil, gas, and minerals surged ahead, often with little regard for long-term consequences. As the global population grew, so too did our reliance on the ocean to meet our needs, from food and transportation to climate regulation. But these assumptions are now proving dangerously misguided.

The effects of this relentless pressure are evident in the ocean's shifting baseline. Marine ecosystems are showing signs of strain: coral reefs, which have existed for hundreds of millions of years, are collapsing at alarming rates; fish stocks are plummeting; and species that once thrived are being displaced or driven to the brink of extinction. The once-immovable forces of oceanic currents and temperature gradients are fluctuating in ways that affect global weather patterns, influencing droughts, floods, and storms. This is not a distant future; these changes are unfolding now, and they will intensify.

However, while the challenges are frightening, the future of the ocean is not solely a story of decline. There is an opportunity—perhaps a necessity—for the ocean to become a symbol of resilience, adaptation, and renewal. Humanity's relationship with the ocean must evolve. We are at a critical juncture where the actions we take in the coming years will shape the ocean for generations. Just as we have been agents of change, we can also be agents of restoration and protection. The future ocean will demand innovation, cooperation, and a profound shift in how we interact with our planet's most vital resource.

New scientific discoveries and technological advances offer hope for mitigating the damage and fostering recovery. Marine conservation strategies, coupled with international cooperation, can turn the tide toward a sustainable future. The key lies in recognizing the interconnectedness of the ocean with all aspects of life on Earth. The ocean is not a separate entity; it is the pulse of the planet, influencing everything from the air we breathe to the climate that sustains us. Understanding its future means understanding our own.

This book sets out to explore the possible futures that await the ocean. By examining the forces driving change, the organisms that will thrive or struggle, and the innovations that can safeguard marine ecosystems, we can begin to paint a picture of what lies ahead. The future ocean will be one of contrasts—where challenges and solutions coexist, where fragility meets resilience, and where the actions of today will ripple through the waters of tomorrow.

I invite readers to embark on this journey into the future of the ocean, not only to understand the risks and opportunities but to appreciate the profound importance of the seas. The story of the ocean is, in many ways, the story of life on Earth. Its future is tied to the choices we make now—choices that will define the world for centuries to come.

Acknowledgments

I deeply thank Thomas Kiørboe for his insightful comments, which have significantly enhanced the quality and depth of this book.

Contents

1 The Ocean in Human Perception — 1
The Illusion of a Clean and Perfect Ocean — 1
Eight Billion Godzillas — 3
The Tragedy of the Commons: Exploiting a Seemingly Endless Resource — 5
The Infinite Trash Can — 9
The Marine Food Web and Its Delicate Balance — 10
Actions Have Consequences — 11
Further Reading — 12

2 Climate Change and Ocean Warming — 15
The Ocean's Role in Climate Regulation — 15
The Gulf Stream and the Slowing of the Atlantic Meridional Overturning Circulation — 17
The Particular Case of Polar Oceans — 19
Increased Stratification and the Role of the Thermocline — 23
Ocean Heatwaves and Their Impact on Marine Life — 24
Species Replacement: The Cod-Calanus Example and Beyond — 25
Commercial Impacts of Species Shifts — 27
Further Reading — 29

3 Sea-Level Rise: A Looming Catastrophe — 31
The Accelerating Rise: Human Influence on Sea Levels — 31
Disproportionate Impacts on Coastal Communities — 33
Coastal Erosion and the Loss of Ecosystems — 35

  Socioeconomic Implications: Disruption to Livelihoods   36
  Solutions and Their Limitations   37
  Recommendations for Moving Forward   38
  Further Reading   39

4 Ocean Acidification: The Silent Crisis   41
  The Ripple Effect on Marine Life   41
  The Fragility of Plankton and the Food Web   42
  Effects of Acidification on Non-calcifying Animals   44
  Acidification as a Compounding Stressor   45
  Long-Term Persistence and Hot Spots of Vulnerability   46
  Socioeconomic Implications   47
  Public Engagement and Awareness   47
  Further Reading   48

5 The Polluted Future of Our Oceans   51
  Classic Pollutants: Oil Spills, Nutrient Runoff, and Heavy Metals   51
  Emergent Pollutants: Microplastics and Chemicals   55
  Eutrophication, Dead Zones, and Ecosystem Collapse   59
  Human Health Impacts: Pollution in Our Food and Water   61
  The Future of the Ocean and Human Health   62
  Further Reading   64

6 Biodiversity Under Threat   67
  The Accelerating Loss of Marine Life   67
  Comparing Biodiversity Loss Across Land and Sea   72
  Restoring Marine Biodiversity: A Path Forward   74
  Further Reading   74

7 The Future of Marine Plankton   77
  Plankton and Climate Regulation: Earth's Hidden Thermostat   77
  Plankton Shifts: Ecosystem Mismatches and Phenological Crises   79
  Expanding Oxygen Minimum Zones   81
  Plankton as a Food Source for an Overpopulated Planet   82
  Challenges and Risks of Plankton-Based Food Production   83
  Further Reading   84

8 The Rise of Jellyfish Dominance — 87
- Resilience of an Ancient Organism — 87
- The Impact of Overfishing: Declining Predators — 88
- Climate Change: Warming Waters and Expanding Habitats — 89
- Pollution: Fueling Jellyfish Blooms — 90
- Impact on Economy and Infrastructures — 90
- Preventing a Jellyfish-Dominated Future — 94
- Further Reading — 95

9 The Consequences of Invasive Species — 97
- The Case of *Mnemiopsis leidyi* — 97
- The Lionfish Invasion: An Expanding Crisis — 100
- The European Green Crab (*Carcinus maenas*) — 103
- *Caulerpa taxifolia* Invasion — 105
- The Role of Aquaculture and Artificial Structures in Species Invasions — 107
- Biocontrol Measures: Benefits and Risks — 109
- The Urgent Need for Prevention and Management — 111
- Further Reading — 114

10 The Future of Fisheries — 115
- The Dual Threat of Overfishing and Climate Change — 115
- The Socioeconomic Divide: Industrial vs. Small-Scale Fisheries — 117
- The Future of Seafood: A Changing Diet — 117
- Is Aquaculture a Sustainable Solution? — 119
- Socioeconomic and Ecological Consequences of Future Fisheries — 122
- Preventing Fisheries Collapse: A Path Forward — 122
- Further Reading — 124

11 The Overexploitation of Sharks — 125
- The Overexploitation of Sharks — 125
- The Ecological Role of Sharks — 127
- Socioeconomic Impacts of Shark Declines — 128
- The Future Without Sharks: Ecosystem Collapse — 129
- What Can Be Done? — 129
- Further Reading — 130

12 Whales: Guardians of Marine Ecosystems and the Global Battle for Their Protection — 131
Whales' Ecological Role — 131
The Threats Facing Whales Today — 132
International Efforts to Protect Whales — 134
The Future of Whales and Marine Ecosystems — 135
Further Reading — 136

13 Deep-Sea Ecosystems and Exploitation — 137
The Unique Deep-Sea Communities Based on Chemical Energy — 137
The Threat of Deep-Sea Mining — 139
Waste Disposal in the Deep Ocean: A Hidden Crisis — 140
The Legal and Regulatory Landscape — 141
Further Reading — 142

14 The Role of Marine Protected Areas — 143
Defining Marine Protected Areas and their Importance — 143
The Ecological Benefits of Marine Protected Areas — 144
Marine Protected Areas and Climate Change — 146
Challenges in Marine Protected Area Management — 148
Connectivity and Technological Advancements in Marine Protected Areas — 149
Integrating Marine Protected Areas into Global Ocean Governance — 150
Further Reading — 151

15 Emerging Technologies — 153
Satellite Monitoring and In Situ Monitoring: A Global View of Ocean Health — 153
Big Data and Predictive Modelling: The Digital Revolution — 156
Balancing Innovation with Traditional Expertise — 157
Some Solutions Ahead: Ocean Cleanup and Carbon Sequestration — 158
Further Reading — 160

16 Future Ocean Resilience and Adaptation Strategies 163
Restoration of Coastal Habitats 163
Species Adaptation and Plasticity 164
Human Adaptation and Community-Based Solutions 168
Further Reading 169

17 Redefining the Human-Ocean Relationship 171
Humanity's Destructive Impact on the Ocean 171
The Illusion of Progress and Infinite Growth 172
Exploitation at the Heart of the Problem 172
Hypocrisy in Ocean Management 173
The Need for Political Will and Education 175
A New Relationship with the Ocean: Stewardship
 and Responsibility 175
A Few Words of Hope 176
Further Reading 177

About the Author

Albert Calbet is a marine biologist at the Institute of Marine Sciences (CSIC) in Barcelona, Spain, where his research focuses on the ecology and ecophysiology of zooplankton. Over the course of his career, Albert has made significant contributions to the understanding of marine food webs, particularly through his work on microzooplankton.

Albert has published over 130 peer-reviewed articles and authored multiple outreach books, dedicated to making marine science accessible to the broader public. His expertise and commitment to education extend to his roles as a mentor and educator, guiding students at the PhD, Masters, and undergraduate levels. He has served as Deputy Director at the Institute of Marine Sciences, and his research has been supported by prestigious institutions. He regularly participates in scientific conferences, acts as a reviewer for funding agencies, and serves on editorial boards of high-impact scientific journals.

In addition to his scientific work, Albert is an advocate for science communication, engaging with the public through social media, articles, conferences, and his outreach books, aimed at inspiring the next generation of marine scientists.

1

The Ocean in Human Perception

Throughout history, humans have held a deep connection with the ocean. It has been seen as a source of food, a highway for exploration, a vast mystery to conquer, and even a spiritual entity. However, despite our reverence for the sea, our perception of it remains largely superficial. We view it as a boundless, resilient body that can absorb our impacts without suffering the visible scars we associate with terrestrial ecosystems. This perception, though deeply ingrained in human consciousness, has led to a dangerous underestimation of the ocean's fragility and the intricate ecosystems it sustains.

When we cut down a forest, the destruction is immediate and visceral. Entire ecosystems are leveled in a matter of hours, leaving behind barren landscapes where once there was vibrant life. The emotional response that this triggers is substantial. People understand that such an act causes widespread harm. The loss of trees, plants, and animals is visible, and the absence of life is striking. But when it comes to the ocean, our actions often do not provoke the same response, despite being just as destructive. The ocean hides its wounds beneath the surface, away from our direct line of sight, which allows us to pretend the consequences are less severe. Yet, the ecosystems within the ocean are just as fragile and interconnected as those on land.

The Illusion of a Clean and Perfect Ocean

Many of us dream of a pristine ocean—one that is crystal blue, teeming with life (Fig. 1.1) and seemingly untouched by human hands. We picture schools of vibrant fish darting through coral reefs, dolphins leaping from the waves,

Fig. 1.1 Underwater picture of a pristine ecosystem. Komodo Island. (Author Albert Calbet)

and turtles gliding gracefully along the ocean floor. It is a comforting image that fills us with a sense of wonder and peace. We enter the water for a swim or a dive, imagining we are visiting an idyllic, untouched world. Yet, this vision of the ocean is deeply flawed. We tend to view the ocean, and indeed the natural world, through a deeply anthropocentric lens. We search for images and narratives that align with our human sense of beauty and wonder, often romanticizing marine life and habitats without fully understanding the broader ecological processes at play. When we observe dolphins or octopuses, we cannot help but project our own cognitive and emotional frameworks onto these animals. We ask whether they are as intelligent as we are or if they exhibit behaviors that mirror our own. These comparisons, however well-intentioned, stem from a flawed assumption that humans are the reference point for all life on Earth.

This perspective limits our understanding of the true complexity of life in the ocean. Life in the ocean is not simply an aesthetic experience for human enjoyment. It exists in ways that are often incomprehensible to us, shaped by millions of years of evolution and adaptation to a wide range of environments, from the sunlit surface waters to the pitch-black depths. As we measure and evaluate marine ecosystems, we prioritize those aspects that appeal to our senses or which align with our expectations of what life should look like. We measure biodiversity by species we can see or count easily, we value ecosystems based on their ability to serve us, and we are fixated on species that exhibit behaviors that resemble our own.

Yet, this human-centered view overlooks the fact that we are merely one of many species coexisting on this planet. Life is ubiquitous and diverse, and it thrives in forms and places far removed from our experience. Microbes in deep-sea hydrothermal vents, for example, survive in conditions that would be utterly hostile to human life, yet these organisms play a fundamental role in regulating the chemistry of the oceans and supporting food chains at the ocean floor. These ecosystems remind us that life can adapt to environments that challenge human understanding and that our value judgments are not universal measures of importance or success. This realization should push us toward a paradigm shift in how we think about our place in the world. We must stop viewing ourselves as the center of the universe and start seeing ourselves as one part of an interconnected system where life thrives in myriad forms, both seen and unseen.

In order to truly comprehend the future of our oceans and ecosystems, we need to shed our anthropocentric perspective. It is time to stop looking at nature through the lens of what pleases or benefits us and start recognizing the inherent value of all life forms and processes, regardless of how they relate to human existence. Only then can we begin to make decisions that truly reflect the needs of the planet as a whole, rather than just our species.

Eight Billion Godzillas

Every time we step into the ocean, even for something as simple as a recreational swim, we are disrupting an ecosystem. The delicate balance of life beneath the waves is constantly shifting in response to our actions. We may not see the consequences immediately, but they are there. Sunscreen washes off our skin, introducing chemicals into the water. Boats drop anchors onto sensitive seafloor habitats, damaging coral and seagrass. Coastal development, aimed at bringing us closer to the ocean, erodes natural habitats like mangroves and marshes. To put this in perspective, imagine Godzilla visiting one of our cities—an enormous creature causing destruction as it stomps through streets and buildings. One Godzilla rampaging through a city for a little while would wreak havoc, but eventually, the damage could be repaired within months or years. Now, imagine eight billion Godzillas, equipped with technology to hunt humans, build in our cities, and permanently reshape the landscape, day after day. The destruction would be unstoppable, with entire ecosystems crushed beyond recognition. This is how humanity behaves in the ocean—not a single, temporary disruption but a relentless, daily invasion. With every boat, every building project, and every drop of chemical runoff, we act like eight billion Godzillas, altering the natural balance and leaving a mark that the ocean may never fully recover from. We must realize that these daily disruptions compound over time, leaving a lasting scar on marine ecosystems.

We do not need to enter into the ocean to harm it. Even the most quotidian actions may have consequences in the marine ecosystem. One example of our everyday impact is the release of microplastics from washing synthetic fabrics like polyester, nylon, and acrylic. Every time we do laundry, millions of microplastic fibers are shed from our clothing and flushed into wastewater systems, eventually making their way into rivers and oceans. A single load of laundry can release anywhere from 700,000 to several million microplastic fibers. These tiny particles are virtually invisible to the naked eye but have a significant cumulative effect. According to estimates, over 35% of the microplastics in the ocean originate from the washing of synthetic textiles. Overall, the present concentrations of microplastics in the ocean is very small; however, their concentrations are increasing, particularly in coastal populated areas. These microplastics may be ingested by some marine organisms, which mistake them for food, leading to physical harm, digestive blockages, and long-term health effects. These pollutants may also act as carriers for toxic chemicals, further exacerbating their impact on marine life.

Our small, everyday activities—from washing clothes to swimming—seem trivial, but when multiplied across billions of people, they result in massive disruptions to marine ecosystems. Even eco-friendly products, like biodegradable sunscreens, can still have unintended consequences on the delicate balance of marine habitats. This realization forces us to confront an uncomfortable truth: every action we take has consequences, even if they are not immediately visible. We are not Godzillas, and our activities may seem harmless in isolation, but collectively, they are adding up to a profound alteration of the natural world. If we wish to preserve the beauty and functionality of our oceans, we must rethink not just how we engage with them recreationally but also how we live and consume in our daily lives. The ocean's resilience is remarkable, but it is not infinite. The more we demand from it without considering the cost, the closer we come to tipping it past the point of recovery.

The Tragedy of the Commons: Exploiting a Seemingly Endless Resource

The concept of the "tragedy of the commons" is a powerful lens through which to understand humanity's unsustainable exploitation of ocean resources. This economic theory, introduced by Garrett Hardin in 1968, describes how individuals, acting in their own self-interest, overexploit shared resources to the point of depletion, even though they are aware that their actions harm the collective good. The ocean, vast and seemingly endless, has long been treated as a common resource that can be exploited without consequence. This perception is deeply rooted in the assumption that because the ocean covers over 70% of the Earth's surface, it is too vast to be significantly impacted by human activities. Unfortunately, this could not be further from the truth. The unchecked exploitation of marine resources, particularly through overfishing, habitat destruction, and pollution, has led to significant ecological damage, with consequences for both marine ecosystems and the human populations that rely on them.

One of the most visible examples of resource overexploitation in the ocean is overfishing (Chap. 10). According to the Food and Agriculture Organization (FAO), over 34% of global fish stocks are now being fished at biologically unsustainable levels, a drastic increase from just 10% in the mid-1970s. This is primarily driven by the high demand for seafood, advances in fishing technology, and inadequate regulation, particularly in international waters where oversight is weak. The Atlantic bluefin tuna (*Thunnus thynnus*; Fig. 1.2) is a prime example of how overfishing has pushed a once abundant species to the brink of extinction. Valued for its high-quality meat, particularly in the sushi market, bluefin tuna populations have plummeted by over 80% in certain regions. This decline has led the species to be classified as critically endangered by the IUCN. Similar stories can be seen in many other species, such as cod in the North Atlantic. The collapse of the Canadian cod fishery in the 1990s is a stark reminder of how quickly a resource can be depleted. For centuries, cod was a staple for coastal communities, but overfishing, driven by modern trawling technologies, led to the collapse of this once-thriving fishery. Despite decades of fishing bans, cod populations have yet to fully recover, demonstrating that the damage caused by overexploitation is often difficult, if not impossible, to reverse.

Another major source of ocean exploitation is deep-sea mining, a growing industry that targets valuable minerals such as manganese, cobalt, nickel, and rare earth elements found in polymetallic nodules, hydrothermal vents, and

Fig. 1.2 Bluefin tuna taken from subsurface waters with longline gear during exploratory fishing. (Image credit: NOAA Central Library Historical Fisheries Collection; P. 365 of Our Changing Fisheries 1970 Centenary NMFS centenary volume)

seamounts. These minerals are crucial for manufacturing modern technologies like smartphones, electric vehicle batteries, and renewable energy systems. While mining on land has long been associated with environmental degradation, deep-sea mining presents an even greater challenge due to the sensitivity of marine ecosystems and the limited understanding of the deep ocean's biodiversity. Companies are increasingly eyeing vast areas of the seabed for extraction, especially in the Pacific's Clarion-Clipperton Zone, where rich mineral deposits exist. The International Seabed Authority (ISA) has already granted exploration licenses to several countries, and commercial operations are expected to ramp up in the coming decades. However, scientists warn that deep-sea mining could cause irreversible damage to fragile ecosystems that have taken millennia to form. Unlike terrestrial ecosystems, deep-sea habitats have slow recovery rates, and disturbances could have lasting impacts on species we have yet to fully understand. Despite these concerns, the demand for minerals continues to rise, pushing the boundaries of ocean exploitation in new and uncertain ways.

A further example of ocean resource exploitation is the harvesting of marine life for products other than food. While fishing for consumption is one of the most well-known uses of marine resources, there is a growing market for nonfood marine products such as cosmetics, pharmaceuticals, and luxury items. For instance, shark populations are heavily exploited (Chap. 11) not only for their meat but also for their fins, which are highly prized in traditional Chinese medicine and cuisine. The practice of shark finning—where sharks are caught, their fins removed, and the rest of the body discarded—has led to the severe decline of shark populations worldwide. Estimates suggest that over 100 million sharks are killed each year, primarily for their fins, contributing to the collapse of many shark species. The loss of apex predators like sharks has cascading effects on the marine food web, altering the balance of entire ecosystems. Similarly, other marine animals, such as sea cucumbers, are harvested for their use in traditional medicines and as delicacies. Sea cucumbers play an important role in nutrient recycling and sediment health on the ocean floor, yet their populations have been depleted in many areas due to overharvesting. In some parts of Asia and the Pacific, illegal and unregulated harvesting of sea cucumbers has become rampant, leading to their local extinction in certain areas. As global demand for such products grows, pressure on marine species will continue to rise, threatening not only biodiversity but also the ecological functions these species perform.

Aquaculture, while often promoted as a solution to overfishing, has also become a major source of resource exploitation, particularly in coastal regions. The global expansion of aquaculture, especially in countries like China, Norway, and Chile, has led to the large-scale farming of species such as salmon, shrimp, and tilapia. While aquaculture can provide a sustainable source of seafood if managed correctly, it also comes with significant environmental costs. Intensive shrimp farming, for example, has led to the destruction of large areas of mangroves and coastal wetlands, which serve as important breeding grounds for many marine species. In addition, the concentration of fish farms in coastal areas can lead to the accumulation of waste, excess nutrients, and chemicals in the surrounding waters, causing eutrophication and the spread of diseases to wild fish populations. The overuse of fishmeal and fish oil as feed in aquaculture further exacerbates pressure on wild fish stocks, as smaller fish species are harvested in large quantities to support the growing aquaculture industry. This creates a paradox where the aquaculture industry, intended to alleviate pressure on wild fish populations, contributes to their decline.

Marine tourism also represents a significant form of ocean exploitation. Coastal and marine environments attract millions of tourists each year, drawn by the allure of pristine beaches, clear waters, and vibrant marine life. While tourism can bring economic benefits to local communities, it often comes at the cost of environmental degradation. Overcrowding in popular coastal destinations leads to the destruction of fragile marine habitats such as seagrass beds, kelp forests, and breeding grounds for various marine species. Furthermore, unregulated tourism activities like snorkeling, diving, and boat tours can disturb wildlife, including marine mammals and seabirds. For example, whale watching, while considered a low-impact activity, can stress cetaceans, altering their feeding and breeding behaviors. In regions where tourism infrastructure is poorly managed, such as many small island nations, the impact on marine ecosystems can be devastating. Without proper regulation and sustainable practices, the pressure from marine tourism will only continue to grow as the global tourism industry expands.

Finally, the pursuit of hydrocarbons through offshore oil and gas drilling represents another significant form of resource exploitation. The discovery of vast oil reserves beneath the seabed has led to a boom in offshore drilling, particularly in regions like the Gulf of Mexico, the North Sea, and the Arctic. While offshore oil drilling has provided energy resources that fuel the global economy, it also carries immense risks to marine ecosystems. Oil spills, such as the Deepwater Horizon disaster in 2010, illustrate the catastrophic effects that accidents in offshore drilling can have on the ocean. The spill released millions of barrels of oil into the Gulf of Mexico, causing widespread damage to marine life and coastal communities. Even routine drilling operations release pollutants, disrupt marine habitats, and contribute to noise pollution that affects marine mammals like whales and dolphins. As the demand for oil continues, companies are pushing further into remote and ecologically sensitive areas, such as the Arctic, where ice melt is opening new areas for exploration. The exploitation of hydrocarbons in these regions raises serious concerns about the long-term impacts on fragile ecosystems that are already under stress from climate change.

In conclusion, the tragedy of the commons in the ocean extends far beyond overfishing and pollution. From deep-sea mining to aquaculture, and from tourism to the extraction of hydrocarbons, humans are exploiting ocean resources in increasingly diverse and unsustainable ways. Each of these activities carries significant environmental risks and, if not managed properly, can lead to the irreversible degradation of marine ecosystems. As the global population continues to grow and demand for resources increases, it is essential that we recognize the limits of the ocean's ability to provide and implement more sustainable practices. The ocean may seem endless, but its resources are finite, and without careful stewardship, the consequences of overexploitation will be felt by generations to come.

The Infinite Trash Can

Human perception often treats the ocean as an infinite trash can, a vast and bottomless reservoir capable of holding all the waste we generate. This mindset has led to a staggering level of pollution (Chap. 5), silently and persistently eroding the health of marine ecosystems. One of the most visible forms of pollution is plastic waste, now found in nearly every part of the ocean, from the surface waters to the deep sea. Every year, between 8 and 12 million metric tons of plastic enter the ocean, where they break down into smaller particles, creating a nearly ubiquitous presence of microplastics. These microplastics are ingested by marine animals—fish, seabirds, and turtles—often mistaken for food, with deadly consequences. For instance, a study in 2020 revealed that 90% of seabird species have ingested plastic, and it is estimated that by 2050, nearly all species will have ingested some form of plastic.

Beyond visible plastics, marine pollution also involves the infiltration of chemical pollutants such as pesticides, heavy metals, and industrial by-products. These contaminants accumulate in the tissues of marine organisms, becoming more concentrated as they move up the food web in a process known as biomagnification. Top predators like sharks, tuna, and whales are especially vulnerable to these chemicals. Studies have shown that mercury levels in top predators can be ten to ten thousand times higher than in the surrounding water. Such concentrations can cause serious health issues, including reproductive failure, developmental disorders, and death.

One example of the invisibility of this destruction is nutrient runoff from agricultural practices. Excess nitrogen and phosphorus from fertilizers enter rivers and eventually flow into the ocean, causing eutrophication—a process that leads to oxygen-deprived "dead zones" where marine life cannot survive. A dramatic example is the Gulf of Mexico's dead zone, which spans up to 6000–7000 square miles each year due to nutrient pollution.

Because many of these pollutants are not immediately visible to the naked eye, the extent of the damage is often overlooked or ignored. However, the consequences are profound. Plastic and chemical pollution not only affect marine species but may also affect humans, particularly as contaminated fish and seafood make their way up the food chain. This presents a clear and present danger to public health and highlights the urgency of addressing marine pollution at its source.

This anthropogenic view of the ocean as an endless repository for our waste needs to be reconsidered. Instead of treating the ocean as an infinite trash can, we need to recognize it as a finite and fragile system, intricately connected to our well-being and the health of the planet.

The Marine Food Web and Its Delicate Balance

To truly understand the consequences of our actions, we must first recognize how intricately connected the ocean's ecosystems are. The marine food web is an intricate and interconnected system that supports life in the ocean, with plankton playing a foundational role. At its base, phytoplankton, microscopic plantlike organisms, drive the primary production in marine ecosystems (Chap. 7). These tiny organisms photosynthesize, using sunlight, carbon dioxide, and nutrients to produce organic matter. They are responsible for producing more than 50% of the planet's oxygen, but most of this oxygen is consumed within the ocean by marine organisms through respiration. Only a small fraction of this oxygen, if any, escapes into the atmosphere.

Phytoplankton are the first step in the marine food web. They are consumed by zooplankton, small animals like copepods and krill that drift in ocean currents. These zooplankton occupy a critical role as primary consumers, transferring the energy produced by phytoplankton to higher trophic levels. For instance, baleen whales feed almost exclusively on krill, consuming tons of these tiny crustaceans daily.

Small fish such as anchovies, sardines, and herring form the next trophic level, feeding on zooplankton and transferring energy up the food chain. These fish are crucial because they serve as prey for larger predatory fish, seabirds, and marine mammals. Their population health is essential for maintaining balance within marine ecosystems. Disruptions at this level, whether from overfishing or habitat destruction, can cause cascading effects that ripple through the entire food web.

At the top of the marine food web are predators like sharks, tuna, and marine mammals like orcas. These apex predators regulate populations of species below them, preventing overpopulation of certain species and maintaining biodiversity. The loss of apex predators due to overfishing or environmental changes can destabilize entire ecosystems, leading to imbalances such as the overgrowth of certain species, which can further affect the food web's dynamics.

Plankton, especially phytoplankton and zooplankton, are not just the base of the food web but also essential to global biogeochemical cycles. Their role in carbon sequestration is critical. Through the biological pump, phytoplankton absorb carbon dioxide from the atmosphere during photosynthesis, which is then transferred through the food web as organisms consume one another. When marine creatures die, their carbon-rich bodies sink to the ocean floor, effectively sequestering carbon for long periods.

Ultimately, the marine food web is a complex and delicate system. Every action we take—whether it is overfishing, pollution, or coastal development—affects this intricate network in ways that can reverberate throughout the ecosystem. The collapse of one species often triggers a cascade of consequences for others, threatening not only marine biodiversity but also the resources upon which human societies depend. Understanding our role in the marine food web is crucial for developing sustainable practices that ensure the long-term health of the ocean and its inhabitants. If we continue to exploit and degrade marine ecosystems at current rates, we risk causing irreversible damage to a system that supports life on Earth.

Actions Have Consequences

At the heart of the issue is how we, as a species, view the ocean. We often believe that if we can see the ocean, clean and beautiful, teeming with life, we are doing well by it. But our actions, both large and small, reverberate through this fragile ecosystem. Every boat trip, every fish farm, or every dive into the water has consequences that we often fail to see or acknowledge. And this leads to a dangerous moral complacency: as long as the ocean looks blue and the fish seem plentiful, we allow ourselves to believe that we are protecting it.

But the truth is, our actions are not without harm. We cannot simply clear our conscience by saying we support conservation efforts or sustainable practices while ignoring the full scope of our impact. This does not mean we should stop using the ocean or enjoying its beauty. On the contrary, the ocean is an invaluable resource, both for our livelihoods and our well-being. It supports millions of people through fisheries, tourism, and countless other industries. The question, then, is not whether we should stop interacting with the ocean. What is important is that we understand we are responsible for what happens in the ocean and act with a consciousness of this responsibility. It is not about avoiding all harm—that is impossible. Every species on Earth has an impact on its environment, and humans are no exception.

What we must do is strive to minimize that harm. This means approaching every decision—whether it is related to aquaculture, fishing, tourism, or recreational activities—with the understanding that our actions carry weight. When we build an aquaculture facility, we must consider not only the economic benefits but also the potential environmental costs and find ways to mitigate them. When we fish, we must do so in a way that allows populations to recover and ecosystems to remain in balance. When we swim in the ocean, we must recognize that even this simple act has an impact and make choices that minimize our footprint.

The ocean is not just a resource to be exploited or a playground to be enjoyed. It is a living, breathing ecosystem that supports life in ways we are only beginning to understand. As humans, we must shift our perception of the ocean from that of an infinite, resilient force to that of a fragile, interconnected system that requires our care and respect.

By adopting this perspective, we can begin to make more informed and responsible choices. We can continue to benefit from the ocean's abundance, but we must do so in a way that acknowledges the consequences of our actions and seeks to minimize harm. The ocean's ecosystems are resilient, but they are not invulnerable. The time has come for us to take responsibility for the impacts we have on the marine environment and to act with the awareness that we are, in many ways, the stewards of the ocean's future. Only then can we ensure that the ocean remains a vibrant and vital part of our world for generations to come.

Further Reading

Duarte, C. M. (2024). *Ocean – The secrets of planet earth* (252 p). Springer Nature. This book provides basic facts for understanding the oceans, their properties, and their importance to mankind through the ages.

Earle, S. (1995). *Sea change: A message of the oceans* (361 p.). G. P. Putnam's. Written by a renowned marine biologist, this book discusses the importance of ocean conservation and the threats posed by human activities.

Helvarg, D. (2006). *Blue frontier: Dispatches from America's ocean wilderness* (2nd ed., 336 p). Sierra Club Books. An insightful look at U.S. marine ecosystems and the challenges they face from industrialization, pollution, and climate change.

Jackson, J. B. C. (2008). Ecological extinction and evolution in the brave new ocean. *Proceedings of the National Academy of Sciences, 105*(Supplement 1), 11458–11465. A scientific article that explains how overfishing, pollution, and habitat destruction are leading to the decline of ocean biodiversity.

Kolbert, E. (2014). *The sixth extinction: An unnatural history* (336 p). Henry Holt. Focuses on the current mass extinction event caused by human activities, including the threats facing marine life.

Moore, C. J. (2012). *Plastic ocean: How a sea captain's chance discovery launched a determined quest to save the oceans* (400 p). Avery. A firsthand account of the discovery of the Great Pacific Garbage Patch and the dangers of plastic pollution to marine ecosystems.

Roberts, C. (2007). *The unnatural history of the sea* (456 p). Island Press. A historical examination of humanity's exploitation of the ocean and its ecosystems, covering key milestones in marine resource depletion.

Roberts, C. (2012). *The ocean of life: The fate of man and the sea* (405 p). Penguin Books. Discusses the significant threats to the world's oceans and marine biodiversity, emphasizing climate change, overfishing, and pollution.

Urbina, I. (2019). *The outlaw ocean: Journeys across the last untamed frontier* (576 p). Knopf. Investigates illegal activities on the high seas, including overfishing and environmental crimes, and their impact on marine ecosystems.

Weis, J. S. (2015). *Marine pollution: What everyone needs to know* (209 p). Oxford University Press. A comprehensive overview of the key pollutants affecting the ocean, such as plastics, chemicals, and agricultural runoff, with a focus on solutions.

2

Climate Change and Ocean Warming

The ocean, covering more than 70% of Earth's surface (Fig. 2.1), has always been a stabilizing force, regulating the planet's climate, weather, and ecosystems. It absorbs much of the sun's energy, buffering temperature fluctuations and supporting life in its many forms. However, over the past century, human activities have altered this balance, leading to rising global temperatures and the warming of the ocean itself. This warming is not just a temporary condition; it signals long-term, potentially irreversible changes that will transform the ocean as we know it.

To understand what lies ahead for the ocean and its creatures, we must acknowledge a fundamental principle of nature and evolution: changes are not fully reversible. Just as an extinct species cannot return, the ocean that existed 100 years ago is beyond recovery. The ocean of the future will be one shaped by new dynamics, driven largely by human behavior. However, this is not a cause for panic or resignation. It is a call to action—an opportunity to shape the future rather than merely react to it.

The Ocean's Role in Climate Regulation

For millions of years, the ocean has acted as Earth's thermal reservoir, absorbing over 90% of the excess heat generated by human activities since the 1970s. This absorption has led to a measurable rise in sea surface temperatures, with the upper 700 meters of the ocean experiencing a temperature increase of 0.11 °C per decade since 1971. In 2021 alone, ocean temperatures hit a record high, continuing a worrying trend. Even seemingly small, these temperature

Fig. 2.1 Earth—Pacific Ocean. This color image of the Earth was obtained by NASA's Galileo spacecraft early Dec. 12, 1990, when the spacecraft was about 1.6 million miles from the Earth. (http://photojournal.jpl.nasa.gov/catalog/PIA00123)

changes are having outsized effects on marine ecosystems. For example, the ocean is warming faster than any other time in the last 11,000 years, which has contributed to more frequent marine heatwaves, with the number of heatwave days increasing by over 54% between 1925 and 2016. Warmer waters lead to the melting of polar ice, changes in species distributions, and the increased stratification of ocean layers, which reduces nutrient mixing and impacts primary productivity, threatening entire marine food webs.

The Gulf Stream and the Slowing of the Atlantic Meridional Overturning Circulation

The potential slowdown of the Atlantic Meridional Overturning Circulation (AMOC) is one of the most pressing global consequences of ocean warming. This vast system of ocean currents plays a pivotal role in regulating the Earth's climate by transporting warm, salty water from the tropics northward to the North Atlantic, where it cools and sinks, driving deep ocean currents. The system acts as a global conveyor belt, redistributing heat and helping to stabilize global weather patterns. However, as polar ice melts and vast amounts of freshwater flow into the North Atlantic—particularly from Greenland's rapidly shrinking ice sheet—the delicate balance that sustains the AMOC is being disrupted. Freshwater is less dense than salty water, making it harder for the water to sink and, as a result, weakening the entire AMOC system.

This process is already well underway. Greenland's ice sheet (Fig. 2.2) is contributing approximately 280 gigatons of ice per year into the North Atlantic, diluting the salinity of surface waters. Studies indicate that AMOC has already weakened by about 15% since the mid-twentieth century, a rate of decline that is unprecedented in over a millennium. Projections suggest that if global temperatures continue to rise at the current pace, the AMOC could

Fig. 2.2 Ice-melting season in Greenland West Coast. (Author Albert Calbet)

weaken by as much as 34–45% by the end of the century or even reach a complete collapse. The implications of this would be catastrophic, affecting not only local climates but also global climate stability.

In Europe, the Gulf Stream, a key component of the AMOC, helps to keep temperatures milder than other regions at similar latitudes. However, a weakened AMOC could bring colder winters, more extreme storms, and other harsh weather patterns. Some models estimate that Northern Europe could experience a temperature drop of up to 5 °C in the event of a significant AMOC slowdown. This would have a far-reaching impact on agriculture, infrastructure, and energy consumption, creating economic challenges and social upheaval. Furthermore, the weakening of the AMOC could result in more frequent and severe weather anomalies, such as prolonged cold spells and intense storms, significantly affecting livelihoods and ecosystems in the region.

The consequences would not be confined to Europe. The eastern seaboard of the United States would face an accelerated rise in sea levels. Without the strong currents of the Gulf Stream pulling water away from the coast, sea levels could rise an additional 10–15 cm (4–6 inches) more than the global average, exacerbating the already severe risks posed by climate-induced sea level rise. This would further endanger coastal cities like New York, Boston, and Miami, which are already facing considerable challenges in defending against rising seas.

On a broader scale, the disruption of the AMOC could affect rainfall patterns across the tropics. Changes in the flow of heat and moisture around the planet could lead to drier conditions in the Amazon and parts of West Africa, significantly affecting ecosystems, agriculture, and food security. Likewise, shifts in monsoon patterns in Asia could cause devastating consequences for millions of people who rely on consistent seasonal rains for their agricultural practices. In a worst-case scenario, AMOC instability could trigger feedback loops that accelerate global climate change, creating further disruptions to weather systems, agriculture, and biodiversity.

Perhaps, most alarming is the potential for a full collapse of the AMOC system, which could have dire consequences. The Younger Dryas period, which occurred roughly 12,000 years ago, saw a rapid cooling of the Northern Hemisphere due to a slowdown in AMOC, resulting in widespread climatic upheaval. While such an event today is less likely, the fact that AMOC is already weakening serves as a reminder of how delicate these systems are and how susceptible they are to the cumulative impacts of climate change.

The ripple effects of a slowing AMOC are vast, influencing everything from sea levels and weather patterns to marine ecosystems and food security. As this vital component of Earth's climate system continues to weaken, the consequences are felt not just regionally, but globally. The potential for more extreme weather events, disruptions to global food supplies, and rising seas underscores the need for urgent action to address climate change and mitigate its impact on the oceans and the planet as a whole.

The Particular Case of Polar Oceans

The polar regions—the Arctic Ocean in the north and the Southern Ocean surrounding Antarctica in the south—are unique and crucial components of Earth's climate system. These cold, ice-dominated oceans are not only home to some of the most specialized ecosystems on the planet, but they also play a pivotal role in regulating global climate. Processes like ice-albedo feedback, carbon sequestration, and deepwater formation, in addition to driving global ocean circulation, are essential to maintaining the stability of Earth's environmental systems.

Both the Arctic and Antarctic oceans host ecosystems finely tuned to the extreme conditions of cold temperatures and seasonal sea ice coverage. In the Arctic, iconic species such as polar bears, walruses, seals, and bowhead whales depend on sea ice for survival, while in the Antarctic, species like emperor penguins, krill, and Weddell seals have evolved similarly within ice-dependent habitats. The ongoing loss of sea ice due to climate change threatens the stability of these ecosystems and disrupts the broader environmental processes that sustain life on Earth.

The Arctic Ocean is warming at an alarming rate. According to the National Snow and Ice Data Center (NSIDC), Arctic Sea ice extent has declined by approximately 13% per decade since satellite monitoring began in 1979. The average thickness of Arctic Sea ice has also decreased by about 50% over the last few decades. In 2020, the Arctic recorded its second-lowest sea ice extent on record, with only 3.74 million square kilometers of ice cover remaining by the end of the summer melt season.

This drastic reduction in sea ice has global implications. The ice-albedo effect, where highly reflective ice is replaced by darker, heat-absorbing ocean water, accelerates regional and global warming. The Arctic is warming at two to four times the global average, a phenomenon known as Arctic amplification. This accelerated warming is already triggering cascading effects, such as altering weather patterns across the Northern Hemisphere, contributing to more frequent and extreme weather events in midlatitudes.

Fig. 2.3 Polar bear in the Arctic. (Author Albert Calbet)

Arctic ecosystems are deeply affected by these changes. Polar bears (*Ursus maritimus*, Fig. 2.3), which rely on sea ice as a hunting platform for seals, are facing increased difficulty in accessing prey, leading to declines in body condition, reproduction, and survival. Recent data suggest that by 2100, polar bear populations in many regions could face near extinction as sea ice continues to diminish.

In addition to marine mammals, Arctic Sea ice is critical for maintaining productive ecosystems at lower trophic levels. Phytoplankton and sea ice algae form the base of the Arctic food web, supporting zooplankton and, ultimately, larger predators. The loss of sea ice is shifting the timing and location of phytoplankton blooms, which in turn disrupts food availability for species like Arctic cod (*Boreogadus saida*) and seabirds. This disruption could have significant effects on Arctic fisheries, which are valued at approximately $1 billion annually, and on the communities that depend on them.

Another key concern is the potential for permafrost thaw, which could release vast quantities of methane—a potent greenhouse gas—from frozen soils. Current estimates suggest that up to 1400 gigatons of carbon could be stored in Arctic permafrost, more than double the amount currently in the atmosphere. As temperatures rise, this permafrost is thawing, raising the possibility of a positive feedback loop that could further exacerbate global warming.

The Southern Ocean, which surrounds Antarctica (Fig. 2.4), also plays a critical role in global climate regulation. The Antarctic Circumpolar Current

Fig. 2.4 The Antarctica. Gerlache Strait. (Author Albert Calbet)

(ACC), the world's largest ocean current, helps isolate the cold waters of Antarctica, preventing warm water from reaching the continent. However, recent studies have shown that the Southern Ocean is absorbing a significant portion of global warming—approximately 75% of the excess heat and up to 40% of the carbon dioxide emitted by human activities. This heat absorption is contributing to the melting of Antarctic ice shelves, particularly in West Antarctica.

In terms of sea ice, Antarctic Sea ice has shown more variability than its Arctic counterpart. For decades, sea ice extent in the Southern Ocean was relatively stable or even slightly increasing, but since 2014, Antarctic Sea ice has been in sharp decline. In 2023, the Antarctic recorded its lowest sea ice extent on record, with just 1.73 million square kilometers of ice remaining by the end of the summer melt season—a decrease of nearly one million square kilometers compared to the average over the last 40 years.

The decline of sea ice has direct consequences for the Southern Ocean food web. Antarctic krill (*Euphausia superba*), which depends on sea ice for feeding on algae during the winter, is a cornerstone species in the Southern Ocean ecosystem. Krill biomass has been declining in key areas of the Southern Ocean, with a drop of up to 80% in some regions over the past 40 years. This decline is already affecting species like the emperor penguin (*Aptenodytes forsteri*), which relies on krill to feed its chicks. A 2021 study projected that over 80% of emperor penguin colonies could be quasi-extinct by 2100 if current sea ice trends continue.

Marine species in Antarctica are adapted to extremely cold, stable environments, making them particularly vulnerable to warming waters. The rise in sea temperatures also threatens to destabilize the Antarctic ice sheets, particularly the West Antarctic Ice Sheet, which holds enough ice to raise global sea levels by over 3 meters. The Pine Island and Thwaites glaciers have shown accelerated melting in recent years, contributing to rising sea levels. If these glaciers collapse, the resulting sea level rise would have catastrophic consequences for coastal cities worldwide.

Beyond climate change, human activities are increasingly encroaching on the polar oceans. The retreat of Arctic Sea ice is opening new shipping routes, such as the Northern Sea Route, which could cut shipping times between Europe and Asia by nearly 40%. This increased traffic, however, brings risks such as oil spills, which are particularly challenging to clean up in icy waters, and noise pollution, which can disturb marine mammals like whales that rely on echolocation for communication and navigation. Shipping traffic in the Arctic increased by 25% between 2013 and 2019, and this trend is expected to continue as sea ice recedes.

Fishing pressure is also intensifying in both polar regions. The Southern Ocean is home to valuable fisheries, including those targeting Antarctic toothfish and Patagonian toothfish, which are sold as Chilean sea bass. Despite conservation efforts under the Convention for the Conservation of Antarctic Marine Living Resources, illegal, unregulated, and unreported fishing remains a significant challenge, particularly in remote areas of the Southern Ocean.

Oil and gas exploration remains a potential threat in the Arctic, despite current moratoriums by some nations. The region is estimated to hold approximately 13% of the world's undiscovered oil and 30% of its undiscovered natural gas. While Arctic drilling presents numerous technical and environmental challenges, the possibility of future exploitation looms as global demand for energy persists. An oil spill in the Arctic would have devastating consequences, given the difficulty of cleaning up spills in icy waters and the slow recovery of polar ecosystems.

Increased Stratification and the Role of the Thermocline

One of the critical effects of ocean warming is the increased stratification of ocean layers, which is becoming more pronounced as surface waters warm due to climate change. Oceans are naturally structured into layers based on temperature and salinity, with the epipelagic zone (surface layer) absorbing sunlight and warming, and the thermocline acting as a transition zone where temperature drops sharply with depth. This structure, vital for nutrient cycling and ecosystem health, is being disrupted by global warming.

As surface waters heat up, they become less dense, intensifying stratification. This hinders vertical mixing, crucial for transporting nutrients from the deep ocean to the surface where phytoplankton, the base of the marine food web, thrive. Increased stratification limits the upwelling of nutrient-rich waters, leading to a 6% global decrease in phytoplankton productivity since the 1980s. This reduction particularly impacts regions like the North Pacific, where some areas have seen a 30% decline in productivity. This drop has cascading effects, impacting zooplankton, fish, marine mammals, and seabirds, which rely on phytoplankton as a food source, endangering global fisheries and food security for millions, especially in coastal communities in developing countries.

Furthermore, the thermocline is deepening, particularly in tropical and subtropical regions, isolating nutrient-poor surface waters from deeper nutrient-rich layers. This trend reduces primary productivity and compresses the habitats of species such as tuna, forcing them to hunt in smaller, deeper areas, which increases competition and stress on their populations.

By the end of the century, projections based on current emissions scenarios (RCP 8.5) suggest that global sea surface temperatures could rise by 2–4 °C, further intensifying stratification and deepening the thermocline. Primary productivity in nutrient-poor areas, like the subtropical gyres, may decrease by 20–30%, which would ripple through the food web, reducing fish biomass and affecting fisheries. In regions like the equatorial Pacific and North Atlantic, where upwelling is vital, weakened vertical mixing could lead to a 5–17% decline in fish biomass by 2100.

The deepening thermocline also affects species adapted to specific temperature and nutrient conditions, leading to habitat compression for predatory fish like tuna and billfish, which are migrating to deeper waters, potentially losing up to 40% of their habitat. This deepening is not just an ecological issue but also a socioeconomic one, as it limits the ocean's ability to sequester carbon through processes like the biological pump. Reduced phytoplankton availability will weaken the ocean's role in absorbing atmospheric CO_2, which could accelerate climate change and reinforce the warming cycle.

Future projections are, however, uncertain. While an intensification and deepening of the thermocline keeps being a solid hypothesis, it has been also suggested that the increased intensity of storms may intensify mixing, therefore allowing more nutrients on the upper layers.

Ocean Heatwaves and Their Impact on Marine Life

Marine heatwaves are among the most alarming and rapidly worsening consequences of ocean warming. These events are characterized by extended periods of abnormally high sea surface temperatures, sometimes lasting for weeks or even months, and they are becoming both more frequent and severe due to climate change. The increased intensity and duration of these marine heatwaves pose significant threats to marine ecosystems, which often have limited ability to adapt to such rapid and extreme temperature changes.

One of the most notable examples of a marine heatwave is "The Blob," which affected the Pacific Ocean from 2013 to 2016. During this event, sea surface temperatures off the North American West Coast rose by as much as 3 °C above average, impacting marine life from Alaska to California. This prolonged heatwave led to mass die-offs of marine species, including fish, sea lions, and seabirds, which were either unable to find food or could not cope with the elevated temperatures. The fishing industry suffered significant losses as fish stocks were displaced or reduced due to habitat disruption.

Similarly, Australia's Great Barrier Reef experienced consecutive marine heatwaves in 2016 and 2017, leading to widespread coral bleaching. The reef, a UNESCO World Heritage site, lost about 50% of its shallow-water corals during this period. Coral bleaching occurs when corals, stressed by elevated temperatures, expel the symbiotic algae (zooxanthellae) that provide them with energy. This leaves the corals more vulnerable to disease and, if the temperatures remain high, leads to widespread coral mortality. Such events not only devastate biodiversity but also threaten the livelihoods of communities that rely on tourism and fisheries associated with coral reef ecosystems.

Marine heatwaves also severely impact the foundational species in marine ecosystems. Krill, which are small crustaceans that form a critical component of marine food webs, are particularly vulnerable to temperature changes. In regions like the Southern Ocean, krill serve as the primary food source for penguins, seals, and whales. During marine heatwaves, krill populations can collapse, causing cascading effects throughout the ecosystem. In 2015, a severe marine heatwave in the North Pacific led to a 70% decline in krill populations, which severely affected humpback whales that rely on them for sustenance.

These extreme temperature events are not limited to isolated regions but are becoming a global phenomenon. The frequency of marine heatwaves has increased by 34% between 1925 and 2016, and their intensity has grown by 17% over the same period. By the end of the century, under the most severe climate scenarios, marine heatwaves could become 50 times more frequent, potentially affecting entire ocean basins.

As the ocean continues to warm, marine heatwaves will likely become one of the most significant drivers of change in marine ecosystems, pushing them past their resilience thresholds and leading to widespread biodiversity loss and ecosystem collapse.

Species Replacement: The Cod-Calanus Example and Beyond

Ocean warming is also driving shifts in species distribution, which is altering marine ecosystems in unexpected and complex ways. One well-studied example of this is the Atlantic cod and its relationship with two species of copepods—*Calanus finmarchicus* (Fig. 2.5) and *Calanus helgolandicus*—in the North Atlantic. Cod, a commercially important fish species, rely heavily on *C. finmarchicus* as a food source. This copepod species thrives in colder waters and has long been a cornerstone of the Arctic and North Atlantic food webs. However, as ocean temperatures have risen, *C. finmarchicus* has been migrating northward in search of cooler waters.

At the same time, *C. helgolandicus*, a warm-water species, has been expanding its range into the areas once dominated by *C. finmarchicus*. While *C. helgolandicus* fills a similar ecological niche, it is less nutritious than its colder-water counterpart. This shift in copepod populations directly affects cod, which do not benefit as much nutritionally from *C. helgolandicus*. As a result, cod populations have been declining, not only because of overfishing but also due to changes in the availability of their preferred prey.

This shift has been documented in several studies, which highlight how the decline in *C. finmarchicus* availability has negatively impacted the growth

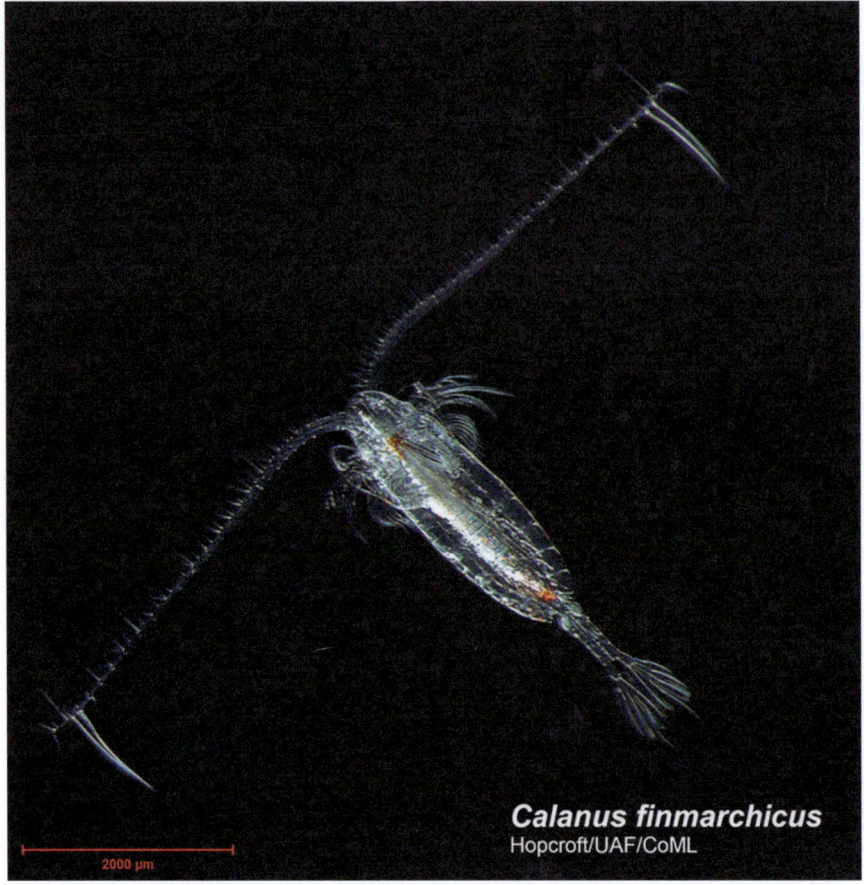

Fig. 2.5 The copepod *Calanus finmarchicus*. (Author R. Hopcroft)

rates and reproductive success of fish species that rely on it. Atlantic cod, a key species in the North Atlantic, has seen significant declines not only due to overfishing but also because of the diminished quality and quantity of its primary food source. The replacement of *C. finmarchicus* by *C. helgolandicus* means that cod populations are not receiving the same energy-dense nutrition, contributing to reduced recruitment and smaller fish sizes. Cod biomass in some areas has decreased by as much as 70% over the past few decades, directly affecting the fishing industry and leading to economic consequences for coastal communities that depend on these fisheries. For example, Norway's cod landings have declined by 30% since 1990, a drop attributed partly to the changing availability of copepods due to warming seas.

The ecological impact of this shift in copepod populations extends beyond fish. Seabirds such as puffins, which feed on small fish like herring and sprats that consume *C. finmarchicus*, have experienced a decline in reproductive success in regions where *C. finmarchicus* is no longer abundant. In some parts of the United Kingdom, seabird populations have decreased by as much as 40% over the last 30 years. This population decline is closely linked to changes in prey availability, illustrating how shifts in plankton communities driven by ocean warming can have cascading effects throughout the food web.

In terms of the broader ecosystem, the decline of *C. finmarchicus* and its replacement by *C. helgolandicus* is creating a ripple effect that is destabilizing food webs. As key species at the base of the food chain are replaced by less nutritious alternatives, the predators that depend on them—ranging from fish to seabirds and marine mammals—are left struggling to adapt. This destabilization is exacerbated by additional stressors like overfishing, pollution, and further climate change, which weaken the resilience of marine ecosystems and make it difficult for populations to recover.

Commercial Impacts of Species Shifts

The economic consequences of species shifts due to ocean warming, such as the replacement of *C. finmarchicus* by less nutritious species like *C. helgolandicus*, or the migration of commercially important fish species like cod, herring, and mackerel, have been significant. In fisheries-dependent nations and coastal communities, these changes are directly impacting livelihoods, national economies, and food security.

For example, in the North Atlantic, where cod stocks have been in steep decline due to warming waters and changing prey availability, the economic impacts have been severe. According to the European Commission, cod fisheries in the North Sea have seen their allowable catches reduced by over 60% between 2003 and 2020 due to declining stock levels, which has resulted in a dramatic reduction in revenue for fishing communities. The financial losses were particularly pronounced in places like Scotland and Norway, where cod is a major component of the fishing industry. The European Union's Scientific, Technical and Economic Committee for Fisheries (STECF) has noted that declining fish stocks have led to substantial income losses, with estimates indicating that some fishing fleets in the North Atlantic lost 30–40% of their annual revenue during this period.

In Iceland, cod has historically been a cornerstone of the economy. A significant downturn in cod populations has forced the country to shift to mackerel and herring fisheries, which are now making up a larger portion of Iceland's fishing revenue. However, the transition has not been smooth. Fisheries management had to invest in new equipment and vessels suited to catching these species, and the shift in the economic structure of fisheries has led to temporary income gaps and unemployment in regions traditionally dependent on cod.

The North Sea's fishing industry has also faced significant economic disruption due to the shifts in species distribution. The mackerel stock, for example, has migrated further north into Icelandic and Faroese waters, sparking diplomatic disputes over fishing rights between the European Union, Iceland, and Norway. This so-called mackerel war has led to economic losses from disrupted trade and negotiations. Mackerel, a valuable species, contributed over €1 billion to the European fisheries economy in 2019, but disputes over fishing quotas and overfishing concerns have put this figure at risk.

On a global scale, the World Bank has estimated that climate change, including ocean warming, could result in a 20% reduction in global fish catch by 2050, which could lead to a loss of around $10 billion annually in revenue for the global fishing industry. This would have a particularly devastating impact on developing countries that rely heavily on fisheries for employment and food security.

Further downstream, the ripple effects of species shifts are also seen in industries such as tourism and food processing. In places where iconic species like cod or specific types of shellfish were once abundant, local economies are suffering from reduced tourism as well as decreased demand from seafood processors and restaurants. For instance, in areas like New England, the collapse of the cod fishery has led to reduced tourism for fishing charters and a decline in seafood-related tourism revenue. Similarly, the loss of traditional species such as cod and haddock has led to increased prices for consumers, affecting the affordability of seafood, especially in countries where fish constitutes a significant portion of the diet.

Moreover, there is clear evidence of an overall poleward migration of many fish species, particularly in the tropics, where the rate of migration has been established at around 26 km per decade. Such displacement would have catastrophic consequences for entire tropical marine ecosystems and would, in turn, significantly impact human populations dependent on these fisheries.

Further Reading

Ditlevsen, P., & Ditlevsen, S. (2023). Warning of a forthcoming collapse of the Atlantic meridional overturning circulation. *Nature Communications, 14*, 4254.

Doney, S. C., Ruckelshaus, M., Duffy, J. E., Barry, J. P., Chan, F., English, C. A., et al. (2012). Climate change impacts on marine ecosystems. *Annual Review of Marine Science, 4*(1), 11–37. Provides a detailed examination of the broad range of climate change effects on marine ecosystems, including temperature changes and ecosystem shifts.

Frölicher, T. L., Fischer, E. M., & Gruber, N. (2018). Marine heatwaves under global warming. *Nature, 560*(7718), 360–364. Discusses the increasing frequency and intensity of marine heatwaves and their impacts on marine ecosystems.

Hansen, J., Sato, M., Ruedy, R., Schmidt, G. A., & Lo, K. (2016). Global temperature change. *Proceedings of the National Academy of Sciences, 103*(39), 14288–14293. Provides an overview of the effects of increasing global temperatures on various Earth systems, including the ocean.

Hoegh-Guldberg, O., & Bruno, J. F. (2010). The impact of climate change on the world's marine ecosystems. *Science, 328*(5985), 1523–1528. A detailed overview of the impacts of climate change on marine biodiversity, including coral reefs and marine heatwaves.

Intergovernmental Panel on Climate Change (IPCC). (2019). *The ocean and cryosphere in a changing climate.* IPCC Special Report. Provides a comprehensive scientific assessment of how climate change is affecting the ocean, including ocean warming, acidification, and changes in marine ecosystems.

Jones, M. C., & Cheung, W. W. L. (2024). Multi-model ensemble projections of climate change effects on global marine biodiversity. *ICES Journal of Marine Science, 72*(3), 741–752.

National Oceanic and Atmospheric Administration (NOAA). (2021). *Global ocean heat content and sea level rise.* NOAA Report. Regularly publishes data on ocean heat content, showing the effects of ocean warming on global sea levels and climate.

Pörtner, H. O., & Peck, M. A. (2010). Climate change effects on fishes and fisheries: Towards a cause-and-effect understanding. *Journal of Fish Biology, 77*(8), 1745–1779. Discusses how increasing ocean temperatures are affecting fish physiology, distribution, and fisheries, particularly focusing on cod and Calanus.

Shepherd, A., Ivins, E. R., & Rignot, E. (2020). Mass balance of the Greenland ice sheet from 1992 to 2018. *Nature, 579*, 233–239. Focuses on the Greenland ice sheet's contribution to sea level rise and its potential impact on the Atlantic Meridional Overturning Circulation (AMOC).

Stroeve, J., & Notz, D. (2018). Changing state of Arctic sea ice across all seasons. *Environmental Research Letters, 13*(10), 103001. Details the decline of Arctic sea ice and its impacts on the broader climate system, including the ice-albedo feedback mechanism.

Thomas, M. K., Kremer, C. T., Klausmeier, C. A., & Litchman, E. (2012). A global pattern of thermal adaptation in marine phytoplankton. *Science, 338*(6110), 1085–1088. This article examines how ocean warming is affecting phytoplankton communities, which play a critical role in marine food webs.

3

Sea-Level Rise: A Looming Catastrophe

As the planet warms, the melting of polar ice and glaciers, combined with the thermal expansion of seawater, is driving one of the most visible and tangible impacts of climate change: rising sea levels. This phenomenon is not just an environmental issue but a socioeconomic crisis that is reshaping coastlines, displacing populations, and redefining the relationship between humans and the ocean. The creeping advance of the sea is already altering ecosystems and threatening the livelihoods of millions of people worldwide, with its impacts likely to intensify over the coming decades.

The Accelerating Rise: Human Influence on Sea Levels

Sea levels have fluctuated throughout geological time scales, driven by natural processes like the advance and retreat of ice ages. However, the current rate of sea-level rise is unprecedented in human history, primarily driven by human-induced global warming. The melting of the Greenland and Antarctic ice sheets, the retreat of glaciers, and the thermal expansion of seawater are combining to create a dramatic rise in sea levels. Since 1880, global sea levels have risen by more than 20 cm, with the rate accelerating significantly in recent decades. Current projections estimate sea levels could rise by between 0.5 and 1 m by the end of the century, depending on the trajectory of global warming.

However, these general predictions do not account for potential tipping points in ice sheet dynamics, which could cause even more rapid and catastrophic sea-level rise. One of the most concerning ice sheets is the Thwaites Glacier in Antarctica, often referred to as the "Doomsday Glacier." This glacier, roughly the size of Florida, is particularly vulnerable due to warm ocean water flowing underneath it, destabilizing its base. Should the Thwaites Glacier collapse, it could contribute over three meters of sea-level rise globally—an outcome that would drastically accelerate the flooding of coastal areas and permanently submerge many regions.

Similarly, Greenland's ice sheet (Fig. 3.1) is undergoing rapid melting. Studies have shown that the rate of ice loss from Greenland has increased sevenfold since the 1990s. If the entire Greenland ice sheet were to melt, it would contribute to a global sea-level rise of about 7 m. While such a scenario is unlikely to play out within this century, the accelerating melt raises concerns about long-term impacts that could stretch well beyond our current planning horizons.

Fig. 3.1 Greenlandic glacier. Godthåbsfjord, west Greenland. (Author Albert Calbet)

Disproportionate Impacts on Coastal Communities

Sea-level rise is already affecting regions across the globe, but its impacts are uneven. Low-lying island nations, such as the Maldives, Tuvalu, and Kiribati, are at the forefront of this crisis. For these nations, sea-level rise is an existential threat. The land that their populations inhabit lies only a few meters above sea level, and even modest increases could render vast portions of their territory uninhabitable. Saltwater intrusion is contaminating freshwater supplies, agricultural productivity is declining, and entire communities are being displaced. The concept of "climate refugees" is no longer theoretical; it is becoming a reality in these regions, posing a challenge to global governance and international law as these populations seek refuge.

Major coastal cities are also facing rising seas, threatening their economic and social fabric. As sea levels continue to rise due to climate change, many coastal cities around the world are facing profound and potentially catastrophic consequences. In the United States, New York and Miami stand on the frontlines of this crisis. New York is already grappling with increased flooding from storm surges, as seen during Hurricane Sandy in 2012, while Miami's porous limestone foundation makes it particularly vulnerable to rising seas and flooding, even on sunny days. In Asia, cities like Jakarta and Tokyo are experiencing similar threats. Jakarta, one of the fastest-sinking cities in the world due to a combination of sea-level rise and land subsidence, has prompted the Indonesian government to plan the relocation of its capital. Tokyo, though protected by sophisticated infrastructure, remains at risk from storm surges and typhoons exacerbated by higher sea levels. Other cities around the world are equally threatened. Venice (Fig. 3.2), Italy, long accustomed to flooding, faces worsening "acqua alta" events that threaten its cultural treasures. The Netherlands is particularly vulnerable to sea-level rise due to its geography, with about one-third of the country lying below sea level and another third situated less than 1 m above it. Historically, the Dutch have developed an extensive system of dikes, dams, and storm surge barriers to protect their low-lying land from flooding. The country is world renowned for its water management expertise, but with rising sea levels, these defenses are facing increasing challenges. Bangkok, Thailand, is sinking under the weight of rapid urbanization and groundwater extraction, with experts warning that large parts of the city could be submerged by 2050. In China, Shanghai's position on the Yangtze River Delta makes it highly vulnerable to

Fig. 3.2 Venice, Italy. (Author Albert Calbet)

storm surges, while Alexandria, Egypt, is contending with coastal erosion that threatens both its infrastructure and its rich archaeological heritage. Cities in developing nations are often the most vulnerable, as seen in Dhaka, Bangladesh, where millions of people could be displaced due to intensified monsoon rains and rising seas, and Lagos, Nigeria, where inadequate drainage systems leave the city prone to severe flooding. Manila, Philippines, is also at high risk, with rising seas and stronger typhoons posing a constant threat to its urban poor living in informal coastal settlements. The threat extends across continents, from the historic heart of Venice to the economic powerhouse of Shanghai, with each city facing unique challenges tied to rising seas, coastal erosion, and extreme weather events. As these cities adapt, the need for global cooperation and sustainable urban planning has never been more urgent, with millions of lives, economies, and cultural legacies hanging in the balance. The economic costs of protecting these cities are staggering, with billions being spent on seawalls, levees, and flood barriers. Despite these investments, such infrastructure is often only a temporary solution, and many regions will ultimately have to face the difficult reality of retreating from the coast.

Coastal Erosion and the Loss of Ecosystems

Coastal erosion, driven by sea-level rise, is compounding the crisis for many communities. Erosion is occurring at a faster rate than ever, fueled by the combination of rising seas, storm surges, and increased wave energy. Coastal erosion threatens not only homes and infrastructure but also ecosystems that play a crucial role in stabilizing shorelines and protecting against storms (Fig. 3.3). Wetlands, mangroves, and salt marshes are nature's barriers against rising seas, acting as buffers that absorb wave energy and prevent flooding.

Unfortunately, these ecosystems are themselves under threat from rising seas. As sea levels rise, many coastal habitats are being squeezed between the

Fig. 3.3 STS-65 Earth observation of Hurricane Emilia in Eastern Pacific Ocean. (NASA)

advancing ocean and human development—what scientists call the "coastal squeeze." Urban expansion and agricultural development leave little room for these ecosystems to migrate inland, leading to their gradual destruction. The loss of these natural defenses weakens coastlines, making them even more vulnerable to storms and further accelerating erosion.

Coral reefs, another vital coastal defense, are also at risk. Reefs not only support marine biodiversity but also act as natural barriers that reduce the energy of waves, protecting shorelines from erosion and storm surges. As sea levels rise, coral reefs face multiple threats. Rising sea temperatures cause coral bleaching, while ocean acidification weakens coral structures, making them more susceptible to physical damage. Additionally, rising sea levels can submerge coral reefs to depths where sunlight is insufficient for photosynthesis, hindering their growth and regeneration. Their degradation would not only lead to a decline in biodiversity but also remove a key buffer protecting coastal communities from storm damage and erosion.

Socioeconomic Implications: Disruption to Livelihoods

Sea-level rise, driven by global warming, poses a significant threat to millions of people, especially those in low-lying coastal and island communities. Coastal areas, home to 680 million people worldwide, are increasingly vulnerable to flooding, storm surges, and erosion. By 2050, projections estimate that 570 coastal cities will face sea-level rise-related risks, potentially displacing over 800 million people by the end of the century if current trends continue.

Industries dependent on coastal stability—fisheries, agriculture, and tourism—are particularly vulnerable to sea-level rise. Coastal tourism alone, which is valued at over $9 trillion annually, is facing significant disruptions. In popular tourist regions such as the Mediterranean and Southeast Asia, beaches are eroding at alarming rates, directly impacting tourism revenues. In the Maldives, for instance, which relies on tourism for 28% of its GDP, rising seas threaten the very existence of beaches and resort infrastructure.

Erosion of shorelines and the increased frequency of storm surges and flooding expose vital infrastructure, including hotels, airports, and restaurants, to damage. This has already led to billions in economic losses globally. For example, Hurricane Sandy in 2012 caused an estimated $65 billion in damage along the US East Coast, much of which was exacerbated by higher sea levels. These disruptions could lead to widespread job losses and economic destabilization in coastal communities heavily reliant on tourism, such as those in the Caribbean, Mediterranean, and Southeast Asia.

Beyond the economic toll, sea-level rise threatens many of the world's most treasured cultural landmarks. Coastal cities such as Venice, renowned for its historic architecture, are increasingly at risk of submersion. In 2019, Venice experienced one of its worst floods in 50 years, highlighting the vulnerability of its cultural sites. Similarly, ancient temples in Thailand, the forts of West Africa, and historic neighborhoods in the United States (such as New Orleans) are at risk of flooding and erosion.

According to the UNESCO World Heritage Center, more than 100 World Heritage Sites are at risk from rising sea levels, including iconic places like Mont-Saint-Michel in France and Robben Island in South Africa. The loss or damage to these landmarks would not only represent a significant cultural tragedy but also result in diminished tourism revenues. Many of these sites are central to the identity of coastal communities, and their destruction would erode cultural and historical continuity, affecting both local economies and heritage.

Solutions and Their Limitations

In response to rising seas, many regions are investing heavily in engineering solutions, such as seawalls, levees, and flood barriers. While these measures provide critical protection for densely populated areas, they are not without limitations. Seawalls can disrupt natural sediment flows, exacerbating erosion in nearby areas and reducing the ability of beaches to replenish naturally. As sea levels continue to rise, the cost of maintaining and upgrading these defenses will increase dramatically, raising concerns about the long-term sustainability of such infrastructure.

Even in the most advanced coastal cities, hard infrastructure solutions may ultimately prove insufficient. As seas rise, the sheer scale of the threat may require more drastic measures, such as the relocation of entire neighborhoods or cities—a strategy known as managed retreat. While politically and socially difficult, managed retreat may be the only viable long-term solution for some regions, particularly those experiencing rapid land subsidence and frequent flooding.

In contrast to hard infrastructure, nature-based solutions offer a more sustainable and holistic approach to addressing sea-level rise. The restoration of coastal ecosystems such as wetlands, mangroves, and coral reefs can enhance coastal resilience by providing natural barriers to flooding and erosion. These ecosystems not only protect shorelines but also support biodiversity and sequester carbon, contributing to climate change mitigation.

Restoring mangroves and wetlands, for example, can create living buffers that reduce wave energy, absorb storm surges, and stabilize coastlines. In addition to their protective benefits, these ecosystems provide habitat for marine species and contribute to local economies through fishing and ecotourism. Coral reef restoration projects, though costly and labor intensive, can help regenerate damaged reefs and protect coastal communities from storm damage.

Recommendations for Moving Forward

In light of these challenges, governments, communities, and individuals must take immediate and coordinated action to mitigate and adapt to the impacts of rising sea levels. The following recommendations offer a way forward:

1. *Accelerate Global Emission Reductions*: Sea-level rise is driven by climate change, and the most effective long-term solution is to reduce greenhouse gas emissions. International cooperation, such as the Paris Agreement, must be strengthened to ensure rapid decarbonization across all sectors.
2. *Invest in Nature-Based Solutions*: Coastal ecosystems such as mangroves, wetlands, and coral reefs are critical in protecting shorelines. Governments should prioritize the restoration and protection of these ecosystems, which offer cost-effective and sustainable defenses against rising seas.
3. *Reconsider Coastal Development*: Urban planning must take rising sea levels into account. New developments should be designed with resilience in mind, and in some cases, relocation away from vulnerable coastlines may be necessary.
4. *Implement Managed Retreat Where Necessary*: For regions where defenses are no longer viable, managed retreat—relocating communities and infrastructure away from coastlines—should be considered. While politically challenging, this approach may be the only sustainable option in the face of accelerating sea-level rise.
5. *Enhance International Support for Vulnerable Nations*: Many of the countries most affected by rising seas are also the least responsible for greenhouse gas emissions. International aid and technical assistance should focus on helping these nations adapt and develop long-term strategies for resilience.

By acting now and implementing these strategies, humanity can better prepare for the realities of a changing coastline, reduce the risks to lives and economies, and build a more resilient relationship with the ocean.

Further Reading

Church, J. A., Clark, P. U., Cazenave, A., Gregory, J. M., Jevrejeva, S., Levermann, A., et al. (2013). Sea level change. In *Climate change 2013: The physical science basis*. Contribution of Working Group I to the Fifth Assessment Report of the Intergovernmental Panel on Climate Change (IPCC). Cambridge University Press. This IPCC report provides a comprehensive overview of the mechanisms driving sea level rise, including glacier and ice sheet melt, thermal expansion, and future projections.

DeConto, R. M., & Pollard, D. (2016). Contribution of Antarctica to past and future sea-level rise. Nature, 531(7596), 591–597. This study models the potential contribution of Antarctica to future sea level rise under different climate scenarios, including the role of tipping points.

Hallegatte, S., Green, C., Nicholls, R. J., & Corfee-Morlot, J. (2013). Future flood losses in major coastal cities. Nature Climate Change, 3(9), 802–806. Analyzes the economic impact of flooding due to sea level rise in major global cities, offering projections and adaptation strategies.

Hauer, M. E., Evans, J. M., & Mishra, D. R. (2016). Millions projected to be at risk from sea-level rise in the continental United States. *Nature* Climate Change, 6(7), 691–695. Estimates the number of people at risk of displacement in the U.S. due to rising seas, focusing on cities like Miami and New York.

Neumann, B., Vafeidis, A. T., Zimmermann, J., & Nicholls, R. J. (2015). Future coastal population growth and exposure to sea-level rise and coastal flooding – A global assessment. PLoS One, 10(3), e0118571. A global analysis of coastal population growth and future exposure to sea level rise and flooding, highlighting vulnerable regions and nations.

Nicholls, R. J., Hinkel, J., Lincke, D., & van der Pol, T. (2019). *Global investment costs for coastal defense through the twenty-first century* (Policy Research Working Paper; No. 8745). World Bank. This working paper examines the economic costs of protecting coastal regions from rising seas, evaluating the effectiveness of hard infrastructure like seawalls and levees.

Oppenheimer, M., Glavovic, B., Hinkel, J., van de Wal, R., Magnan, A. K., Abd-Elgawad, A., et al. (2019). Sea level rise and implications for low-lying islands, coasts, and communities. In *The ocean and cryosphere in a changing climate*. IPCC Special Report. This chapter from the IPCC Special Report focuses on the risks posed by sea level rise to coastal communities and low-lying island nations, offering global and regional projections.

Scambos, T. A., Bohlander, J. A., Shuman, C. A., & Skvarca, P. (2004). Glacier acceleration and thinning after ice shelf collapse in the Larsen B embayment, Antarctica. Geophysical Research Letters, 31(18), L18402. Focuses on the rapid ice loss following the collapse of the Larsen B ice shelf in Antarctica, providing insight into how ice sheet destabilization can contribute to sea level rise.

Spalding, M., McIvor, A., Tonneijck, F. H., Tol, S., & van Eijk, P. (2014). Mangroves for coastal defence: Guidelines for coastal managers and policy makers. In *Wetlands international and the nature conservancy* (42 p). This report highlights the benefits of using mangroves as nature-based solutions to mitigate coastal erosion and storm surge impacts.

Vitousek, S., Barnard, P. L., Fletcher, C. H., Frazer, N., Erikson, L., & Storlazzi, C. D. (2017). Doubling of coastal flooding frequency within decades due to sea-level rise. Scientific Reports, 7(1), 1399. Projects the increasing frequency of coastal flooding events as sea levels rise, with an emphasis on the economic and social consequences.

4

Ocean Acidification: The Silent Crisis

The ocean has long been regarded as Earth's great stabilizer, absorbing vast amounts of carbon dioxide (CO_2) from the atmosphere and helping regulate the planet's climate. Over the past century, the ocean has absorbed nearly one-third of the CO_2 emitted by human activities, reducing the greenhouse effect and delaying the worst impacts of climate change. However, this service comes at a high cost. The increasing levels of CO_2 absorbed by the ocean are triggering a chemical transformation that has profound consequences for marine life. This process, known as ocean acidification, represents a silent crisis unfolding beneath the waves—one that threatens to reshape marine ecosystems and alter the very fabric of life in the ocean.

Ocean acidification is a complex but well-understood process. When CO_2 dissolves in seawater, it forms carbonic acid, which dissociates into hydrogen ions and bicarbonate. The increase in hydrogen ions lowers the pH of seawater, making it more acidic. Since the Industrial Revolution, the ocean's average pH has dropped by approximately 0.1 units—a seemingly small change, but on the logarithmic pH scale, this represents a roughly 30% increase in acidity. This chemical shift is happening at a pace unprecedented in Earth's history, faster than any known acidification event in the past 300 million years.

The Ripple Effect on Marine Life

The most direct victims of ocean acidification are calcifying organisms—those that rely on calcium carbonate to build their shells and skeletons, such as corals, mollusks, and certain species of plankton. As the ocean becomes more acidic, the availability of carbonate ions decreases, making it difficult for these organisms to form and maintain their calcium carbonate structures. Without sufficient carbonate ions, shells become thinner, weaker, and more susceptible to dissolution. For some species, the inability to form shells can be catastrophic, leading to population declines and even local extinctions.

The collapse of calcifying organisms has cascading effects throughout the marine food web. Coral reefs, for instance, are biodiversity hot spots. The process of reef formation relies on the ability of coral polyps to deposit calcium carbonate over millennia. As acidification impairs coral calcification, reef growth slows, and the structural integrity of these ecosystems weakens. This not only threatens the survival of coral species but also endangers the myriad organisms that depend on reefs for food, shelter, and breeding grounds.

Coral reefs are already under assault from rising sea temperatures and pollution, but acidification amplifies these existing threats. The degradation of coral reefs jeopardizes the livelihoods of millions of people who depend on them for fisheries, tourism, and coastal protection. As reefs erode, coastal communities become more vulnerable to storm surges and erosion, highlighting the interconnectedness between the health of marine ecosystems and human societies.

The Fragility of Plankton and the Food Web

Plankton, the foundation of the marine food web, is also directly affected by acidification. Calcifying plankton species, such as foraminifera, coccolithophores, and pteropods (Fig. 4.1), rely on calcium carbonate to form their shells and skeletons. As ocean acidification progresses, it lowers the pH, making it more difficult for these organisms to access the carbonate ions needed to maintain their structures. This weakening of shells makes them more vulnerable to predation and environmental stress, leading to population declines.

For instance, pteropods, often referred to as "sea butterflies," are a vital food source for many larger marine animals, including fish like salmon and seabirds. A study by the National Oceanic and Atmospheric Administration (NOAA) found that in parts of the Southern Ocean, up to 50% of pteropods were suffering from shell dissolution due to acidification. In the North Pacific, similar conditions threaten populations, with widespread damage observed in pteropod shells exposed to more acidic waters.

The decline in calcifying plankton has cascading effects. In the case of foraminifera and coccolithophores, these organisms are integral to the marine carbon cycle because their calcium carbonate shells sequester carbon from the

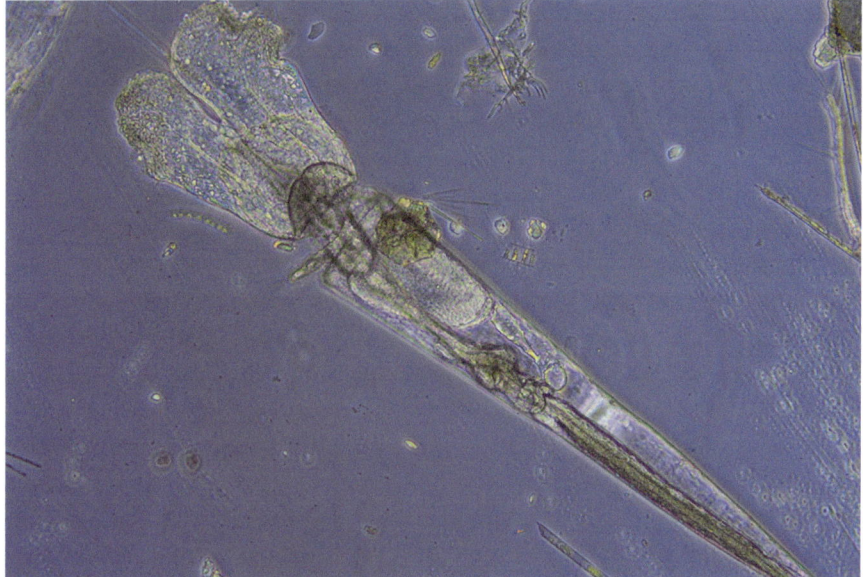

Fig. 4.1 Pteropod. NW Mediterranean. (Author Albert Calbet)

atmosphere. A reduction in their populations not only impacts the food web but also weakens the ocean's role in carbon sequestration, potentially exacerbating climate change.

But the effects of acidification extend beyond calcifying organisms. Non-calcifying plankton, such as diatoms and certain species of zooplankton, are also impacted by changes in water chemistry. While they do not rely on calcium carbonate, studies have shown that changes in pH can affect their growth rates, reproduction, and ability to absorb essential nutrients. For instance, by the year 2200, ocean acidification could lead to a reduction in the dissolution rate of silica from sinking particles, which would decrease the availability of silicic acid in surface waters. This scarcity of silicic acid is expected to trigger a global decline in diatom populations by approximately 13–26%, which could result in significant reductions in primary productivity—the process by which phytoplankton convert sunlight into energy through photosynthesis. Since phytoplankton form the base of the marine food chain, any reduction in their productivity would ripple throughout marine ecosystems, affecting species at all trophic levels.

A decrease in plankton populations threatens biodiversity throughout the marine food web. For example, juvenile fish, which rely heavily on plankton for food, could experience higher mortality rates. Recent estimates show that 10–20% of fish larvae rely directly on plankton like pteropods for their development. Declines in these critical prey species could lead to smaller fish populations, which in turn affect larger predators like tuna, sharks, and whales. Additionally, species such as krill, which are another key food source for marine mammals, could be affected by changes in phytoplankton availability, altering migration patterns and ecosystem dynamics.

The potential collapse of plankton populations poses a grave risk not only to marine biodiversity but also to global food security. Fish and shellfish provide protein to over 3 billion people worldwide, and disruptions to the marine food web could severely impact global fisheries. Coastal communities that depend on fishing for their livelihoods would be particularly vulnerable to these changes, with the economic fallout extending beyond the environmental impacts.

In conclusion, plankton species, both calcifying and non-calcifying, are deeply intertwined with the health of marine ecosystems and the broader climate system. Ocean acidification represents a serious threat to these foundational organisms and, by extension, to marine biodiversity, food security, and the ocean's ability to act as a carbon sink. The scientific data clearly demonstrate the urgency of addressing the root causes of acidification to protect the ocean's intricate food web and the myriad life forms that depend on it.

Effects of Acidification on Non-calcifying Animals

Ocean acidification impacts not only calcifying organisms but also larger non-calcifying marine animals in ways that are complex and far-reaching. One of the key effects on these animals is respiratory stress. Many species, including fish and cephalopods like squid, rely on stable pH levels in seawater to efficiently extract oxygen from their environment. As ocean acidity increases, the balance of ions in the water is disrupted, making it more difficult for animals to regulate their internal pH. This reduced ability to manage pH affects their capacity to breathe, reducing oxygen uptake and the efficiency of their respiratory systems. In particular, squid, which have a unique blood chemistry optimized for efficient oxygen transport, are highly vulnerable to these changes. Studies have demonstrated that even slight decreases in pH can impair the oxygen-binding capacity of their blood, forcing them to expend more energy just to survive. In the long run, this compromises their ability to grow, reproduce, and escape predators, threatening entire populations.

Beyond physiological stress, ocean acidification also triggers behavioral changes in non-calcifying species, especially fish. Studies have shown that acidified waters can interfere with the nervous systems of fish, altering their sense of smell, sight, and even their ability to detect predators. For example, clown fish raised in more acidic conditions lose the ability to detect chemical cues from predators, making them more vulnerable. These behavioral disruptions are not limited to predator-prey interactions but can also affect social behaviors, foraging efficiency, and habitat selection. In more acidic waters, fish may become disoriented and fail to recognize safe habitats, which puts them at risk from both predators and environmental stresses. Such changes in behavior can have cascading effects throughout the food web, particularly in ecosystems where these species serve as prey for larger predators.

Reproductive impacts are another major consequence of acidification on non-calcifying species. Fish and cephalopods experience reduced reproductive success in more acidic environments, a factor that threatens population sustainability. Acidified waters have been shown to impair sperm motility and egg fertilization in some fish species, while others exhibit reduced spawning activity altogether. The effects are particularly concerning for species that play critical roles in both marine food webs and commercial fisheries, such as tuna and cod. Lower reproductive rates can lead to population declines, which, over time, disrupt entire ecosystems and have economic consequences for coastal communities reliant on fishing industries.

Acidification as a Compounding Stressor

Ocean acidification does not occur in isolation. It interacts with other environmental stressors, such as ocean warming, pollution, and overfishing, creating a perfect storm of threats to marine ecosystems. Rising sea temperatures, for example, weaken the resilience of many marine organisms, making them more vulnerable to the effects of acidification. Acidification reduces the ability of some fish to detect predators, making them more susceptible to predation. In others, it affects the sensory capabilities of species like clown fish, altering their ability to navigate and find suitable habitats.

Acidification also affects the ocean's ability to absorb CO_2. As the ocean becomes more acidic, its capacity to act as a carbon sink diminishes, potentially accelerating the pace of global warming. This creates a feedback loop: the more CO_2 we emit, the less the ocean can help mitigate its impact, leading to faster and more severe climate change. Additionally, acidification could exacerbate hypoxic zones—areas of the ocean with low oxygen levels—by altering the balance of chemical processes that sustain these oxygen-depleted environments.

Long-Term Persistence and Hot Spots of Vulnerability

One of the most alarming aspects of ocean acidification is its long-term persistence. Even if humanity were to halt all CO_2 emissions today, the process of acidification would continue for decades, if not centuries, as the ocean slowly absorbs and redistributes the excess carbon already present in the atmosphere. The delayed response of the ocean means that the full effects of acidification have yet to be realized, and the impacts we are seeing now are just the beginning.

Certain regions of the ocean are particularly vulnerable to acidification. Polar waters, for example, are acidifying at a faster rate than tropical or temperate regions because colder water can absorb more CO_2. This makes the Arctic (Fig. 4.2) and Southern Oceans "hot spots" of acidification, where the impacts on marine life are likely to be felt first and most acutely. Upwelling zones, where deep, CO_2-rich waters rise to the surface, are also at higher risk. These areas are critical for global fisheries, and the acceleration of acidification in these regions poses a direct threat to food security.

Fig. 4.2 High Arctic waters during early summer. (Author Albert Calbet)

Socioeconomic Implications

The biological consequences of ocean acidification extend to the human realm, with significant socioeconomic implications. Shellfish industries, for example, are already feeling the effects. In the Pacific Northwest of the United States, oyster hatcheries have experienced mass die-offs linked to acidifying waters. These events threaten the livelihoods of coastal communities dependent on shellfish farming, as well as the broader seafood industry.

Fisheries, which provide food for billions of people worldwide, are also at risk. As acidification alters marine ecosystems, fish populations could decline or shift in unpredictable ways, leading to reductions in catch sizes and economic instability for communities reliant on fishing. The potential for increased competition over dwindling resources may further exacerbate tensions in already vulnerable regions, leading to food insecurity and conflict.

Ocean acidification represents one of the most significant environmental challenges of our time, yet it remains largely invisible to the public. This "silent crisis" requires urgent and coordinated action at both local and global levels. At its core, the solution lies in reducing CO_2 emissions. As the primary driver of both climate change and ocean acidification, cutting greenhouse gas emissions is essential to slowing these destructive processes. International agreements, such as the Paris Agreement, provide a framework for reducing emissions, but progress has been slow, and much more must be done. Restoring habitats that act as natural carbon sinks, such as mangroves, seagrass meadows, and salt marshes, can help buffer against acidification while providing critical habitat for marine species.

Scientific research and monitoring are also crucial. Continued investigation into the mechanisms of acidification, its effects on different species and ecosystems, and its interactions with other stressors will guide the development of effective conservation strategies. This research will help identify regions most at risk and inform policies aimed at mitigating the impacts of acidification.

Public Engagement and Awareness

Perhaps, the greatest challenge in addressing ocean acidification is the lack of public awareness. Despite its profound implications, many people remain unaware of what acidification is or why it matters. This lack of understanding hampers efforts to galvanize political and social action. Public education and outreach will be key to raising awareness and inspiring the behavioral changes necessary to reduce emissions and protect marine ecosystems.

By making the "silent crisis" visible—through media, education campaigns, and community engagement—we can build momentum for the policies and actions needed to address acidification. Understanding the chemical changes taking place beneath the ocean's surface is the first step toward advocating for the protection of this critical resource.

Further Reading

Branch, T. A., DeJoseph, B. M., Ray, L. J., & Wagner, C. A. (2013). Impacts of ocean acidification on marine seafood production: Adaptation options and research needs. *Trends in Ecology & Evolution, 28*(3), 178–186. This paper examines the potential impacts of acidification on global seafood production and outlines possible adaptation strategies for fisheries.

Caldeira, K., & Wickett, M. E. (2003). Anthropogenic carbon and ocean pH. *Nature, 425*(6956), 365. One of the seminal papers that first raised awareness of ocean acidification as a consequence of CO_2 emissions.

Doney, S. C., Fabry, V. J., Feely, R. A., & Kleypas, J. A. (2009). Ocean acidification: The other CO_2 problem. *Annual Review of Marine Science, 1*(1), 169–192. This review covers the science of ocean acidification, its impact on marine life, and how it interacts with other environmental stressors.

Duarte, C. M., Hendriks, I. E., Moore, T. S., Olsen, Y. S., Steckbauer, A., Ramajo, L., Carstensen, J., Trotter, J. A., & McCulloch, M. (2013). Is ocean acidification an open-ocean syndrome? Understanding anthropogenic impacts on seawater pH. *Estuaries and Coasts, 36*(2), 221–236. This paper addresses the geographic variability in acidification and discusses whether it is predominantly a coastal or open-ocean issue.

Fabry, V. J., Seibel, B. A., Feely, R. A., & Orr, J. C. (2008). Impacts of ocean acidification on marine fauna and ecosystem processes. *ICES Journal of Marine Science, 65*(3), 414–432. An influential paper discussing how ocean acidification affects marine fauna, particularly through impacts on physiology and ecosystem processes.

Feely, R. A., Sabine, C. L., Hernandez-Ayon, J. M., Ianson, D., & Hales, B. (2008). Evidence for upwelling of corrosive "acidified" water onto the continental shelf. *Science, 320*(5882), 1490–1492. A study showing how acidified waters from deep upwelling zones affect coastal ecosystems, particularly on the U.S. West Coast.

Gattuso, J. P., & Hansson, L. (Eds.). (2011). *Ocean acidification* (326 p). Oxford University Press. A comprehensive textbook detailing the processes of ocean acidification and its biological, ecological, and socio-economic implications.

Hofmann, G. E., Barry, J. P., Edmunds, P. J., Gates, R. D., Hutchins, D. A., Klinger, T., & Sewell, M. A. (2010). The effect of ocean acidification on calcifying organisms in marine ecosystems: An organism-to-ecosystem perspective. *Annual Review of Ecology, Evolution, and Systematics, 41*, 127–147. This review focuses on the biological impacts of acidification, particularly on calcifying organisms such as corals, mollusks, and plankton.

Kroeker, K. J., Kordas, R. L., Crim, R., Hendriks, I. E., Ramajo, L., Singh, G. S., Duarte, C. M., & Gattuso, J. P. (2013). Impacts of ocean acidification on marine organisms: Quantifying sensitivities and interaction with warming. *Global Change Biology, 19*(6), 1884–1896. An important meta-analysis that quantifies the sensitivity of different marine organisms to ocean acidification and its interactions with ocean warming.

Orr, J. C., Fabry, V. J., Aumont, O., Bopp, L., Doney, S. C., Feely, R. A., et al. (2005). Anthropogenic ocean acidification over the twenty-first century and its impact on calcifying organisms. *Nature, 437*(7059), 681–686. A foundational study using models to project the future of ocean acidification and its effects on marine life.

5

The Polluted Future of Our Oceans

The ocean, once regarded as a boundless expanse of life and vitality, is now a repository for a wide range of pollutants generated by human activities. Pollution in the ocean takes many forms, from plastics to chemical contaminants, from oil spills to nutrient runoff. These pollutants threaten not only marine ecosystems but also human health, making ocean pollution one of the defining environmental challenges of our time. While the most visible form of pollution is plastic waste, the ocean is facing a multitude of chemical, industrial, and agricultural pollutants that are having profound effects on marine life and the broader food web, which ultimately impacts human populations.

Classic Pollutants: Oil Spills, Nutrient Runoff, and Heavy Metals

Among the most infamous forms of pollution are oil spills, which devastate marine ecosystems in both the short and long term. The 2010 Deepwater Horizon disaster, one of the largest oil spills in history, released approximately 4.9 million barrels of oil into the Gulf of Mexico, contaminating over 1000 miles of coastline and killing thousands of marine animals, including dolphins, turtles, and seabirds. Decades earlier, the Exxon Valdez spill in 1989 discharged 11 million gallons of crude oil into Alaska's Prince William Sound, with long-lasting effects that are still being felt today. Even now, some of the affected areas show persistent oil residues, and species like Pacific herring and

Fig. 5.1 Shoreline of Jurong Island, Singapore, showing the multiple oils refineries and industries. (Author Albert Calbet)

sea otters are still suffering population declines as a result of chronic oil exposure.

Oil spills are catastrophic events, but less visible chronic pollution also occurs from operational discharges (Fig. 5.1), including bilge dumping and leaks from drilling platforms. Oil pollution not only coats marine life, leading to suffocation, poisoning, and loss of insulation for animals like otters and seabirds, but it also causes long-term damage to marine ecosystems. Oil toxins accumulate in the tissues of fish and other sea life, which can lead to bioaccumulation and eventually enter the human food chain. Studies have shown that exposure to oil-derived hydrocarbons can cause liver damage, reproductive failure, and cancer in marine species, with potential knock-on effects for humans who consume contaminated seafood.

Nutrient pollution, primarily from agricultural runoff rich in nitrogen and phosphorus, is another classic form of ocean pollution that is causing severe damage to marine ecosystems. Excess nutrients lead to eutrophication, a process where nutrient overloading fuels massive algal blooms. These blooms consume dissolved oxygen as they decay, creating dead zones—areas in the ocean where oxygen levels are so low that most marine life cannot survive. The

Gulf of Mexico dead zone is one of the largest in the world, reaching a peak size of 22,720 square kilometers in 2017. The consequences are profound: fish and shrimp are forced to flee these areas, while benthic (seafloor) organisms that cannot escape often die off in mass events.

The effects of nutrient pollution ripple throughout the food web, disrupting commercial fisheries and collapsing local ecosystems. In many coastal areas, coral reefs, seagrass beds, and mangroves—all critical nurseries for marine life—are suffering from the consequences of nutrient overload, weakening their resilience to other stressors like climate change and ocean acidification.

In addition to oil and nutrient pollution, heavy metals like mercury, lead, and cadmium are being introduced into the ocean through industrial discharges, mining operations, and atmospheric deposition. Mercury is one of the most pervasive and dangerous pollutants. Released primarily from coal-burning power plants, mining, and industrial processes, mercury enters waterways and eventually finds its way into marine environments, where it undergoes conversion into methylmercury, a highly toxic form. Methylmercury bioaccumulates in marine food chains, with concentrations increasing up to 10 million times as it moves from plankton to small fish and ultimately to large predatory fish such as tuna, swordfish, and sharks. For example, in the United States, studies have found that 82% of sampled tuna contained mercury levels that exceeded safe limits for frequent consumption. Similarly, swordfish have been found to contain mercury concentrations of over 0.9 parts per million (ppm), which is far above the 0.3 ppm threshold set by environmental agencies. The risk of methylmercury poisoning is particularly acute for pregnant women, nursing mothers, and young children, as it can impair cognitive development, lead to learning disabilities, and affect motor skills. In fact, the World Health Organization has classified mercury as one of the top ten chemicals of major public health concern.

Lead pollution, another serious issue, often comes from industrial runoff, old piping systems, and atmospheric deposition from leaded gasoline and paints. Once in the marine environment, lead can persist in the water column and sediments, posing risks to both marine life and humans. Lead is known to interfere with the reproductive systems of marine organisms and cause developmental abnormalities. In humans, lead exposure through contaminated seafood can result in a range of health problems, including cognitive impairment and cardiovascular issues.

In addition to mercury and lead, cadmium is another toxic heavy metal posing significant threats to marine ecosystems and human health. Cadmium is introduced into marine environments primarily through industrial activities such as mining, smelting, and the manufacturing of batteries, as well as through agricultural runoff containing phosphate fertilizers. Once cadmium enters the ocean, it can accumulate in sediments and be taken up by marine organisms, particularly filter-feeding species such as mussels, oysters, and clams, which are popular seafood choices in many coastal communities. Research has shown that cadmium can accumulate in the tissues of these organisms, reaching concentrations 100 to 1000 times higher than those found in surrounding waters. In some heavily polluted areas, cadmium levels in shellfish have been measured at levels that exceed safe consumption thresholds set by health agencies. For example, in areas of China's Bohai Sea, cadmium concentrations in oysters have been recorded at 2.6 milligrams per kilogram, far surpassing the 0.5 mg/kg safety limit established by the European Food Safety Authority (EFSA).

The health risks associated with cadmium exposure are serious, particularly for humans consuming contaminated seafood. Cadmium is known to cause kidney damage, weakens bones (leading to conditions like osteomalacia), and is classified as a Group 1 carcinogen by the International Agency for Research on Cancer. Long-term exposure to cadmium through the food chain has been linked to chronic diseases, and in extreme cases, it can lead to Itai-Itai disease, a painful skeletal disorder that was first identified in Japan in the twentieth century due to heavy metal contamination in rice paddies and rivers.

Cadmium pollution, like mercury and lead, is persistent and difficult to remove from the environment. It binds to sediments and can be re-released into the water column through disturbance or natural processes, perpetuating the contamination cycle. This highlights the importance of better regulating industrial waste disposal, reducing agricultural runoff, and monitoring seafood for heavy metal contamination to protect both marine life and human health.

Heavy metals also accumulate in areas with high industrial activity or runoff from urban centers. For instance, regions such as the Mediterranean Sea and Baltic Sea have been identified as hot spots for heavy metal contamination, exacerbated by their semi-enclosed nature, which limits the dispersal of pollutants. In the Baltic Sea, concentrations of cadmium and lead in sediments have been found to be three to ten times higher than in open ocean regions, leading to serious concerns about the long-term impacts on marine life and the food chain.

Furthermore, as these metals are slow to degrade, they persist in the marine environment, continuously cycling through organisms and ecosystems. This not only impacts marine biodiversity but also threatens the safety of seafood that millions of people depend on, creating a global health and ecological crisis that underscores the urgent need for stricter regulations on industrial discharges and better monitoring of pollutants.

Emergent Pollutants: Microplastics and Chemicals

While classic pollutants like oil, nutrients, and heavy metals have long been recognized, emergent pollutants such as plastics, microplastics, and chemical additives have only recently been understood in their full scope. Every year, between 8 and 12 million metric tons of plastic enter the ocean (Fig. 5.2), which is equivalent to roughly one garbage truck full of plastic being dumped into the ocean every minute. Plastics are now ubiquitous in marine environments, from the Arctic to the deepest ocean trenches. Plastic accumulation in ocean gyres, often referred to as "garbage patches," represents one of the most visible consequences of marine pollution. The most well-known of these, the Great Pacific Garbage Patch, is located between Hawaii and California and spans an estimated 1.6 million square kilometers, roughly twice the size of Texas. It is estimated to contain around 80,000 metric tons of plastic debris, with some sources indicating it could hold up to 1.8 trillion pieces of plastic. These gyres form due to the convergence of ocean currents that trap floating debris, which consists primarily of microplastics but also includes larger items like fishing nets, bottles, and other plastic waste. The Great Pacific Garbage

Fig. 5.2 Barcelona coast after a storm showing the plastics and other items left at the beach

Patch alone is growing rapidly, with research suggesting that plastic pollution in the gyres could triple by 2050 if current rates of plastic disposal and production continue unchecked. The harmful effects of plastic pollution within these gyres highlight the urgent need for global efforts to reduce plastic waste, improve waste management systems, and develop biodegradable alternatives.

The plastic debris breaks down slowly under sunlight but never fully biodegrades, fragmenting into smaller and smaller pieces. Microplastics—fragments smaller than 5 millimeters that are pervasive in the water column and sediments— make up about 94% of the estimated 1.8 trillion pieces and may be particularly harmful because they could be ingested by marine organisms at certain levels of the food web. The problem extends to all five major subtropical gyres in the world's oceans, where plastic is accumulating in massive quantities. However, we still lack of global estimates of the in situ harmful effects of these plastics. For instance, recent literature suggests that at the actual concentrations microplastics rarely will be consumed by planktonic organisms.

Marine organisms ingest plastics and microplastics, mistaking them for food, which can have devastating physical and chemical consequences. These plastic fragments often lead to digestive blockages, preventing marine animals from properly processing food, which can result in starvation and malnutrition. For example, sea turtles frequently consume plastic bags, mistaking them for jellyfish, while seabirds often ingest plastic debris floating on the water's surface. Such ingestion can not only obstruct their digestive tracts but also reduce their energy reserves, leading to a higher mortality rate.

Beyond the physical damage caused by ingesting microplastics, the chemical threats associated with plastics are equally alarming. Microplastics may serve as vectors for a wide range of harmful chemicals, including persistent organic pollutants (POPs), heavy metals, and various plastic additives. POPs, which include substances like polychlorinated biphenyls (PCBs) and DDT (which long name is dichlorodiphenyltrichloroethane), adhere to the surface of plastics in seawater, making microplastics a toxic delivery system for marine organisms. These chemicals are known to bioaccumulate through the food web, affecting not only marine animals but also humans who consume seafood.

Plastic additives, such as plasticizers and flame retardants, also pose significant potential threats to marine life. Phthalates and bisphenol A (BPA), two common plasticizers used to enhance the flexibility of plastics, are potent endocrine disruptors. Once ingested by marine organisms, these chemicals interfere with their hormonal systems, leading to reproductive and developmental issues. For example, studies have shown that phthalates can reduce reproductive success in fish, while BPA has been linked to altered sex ratios and impaired embryo development in various marine species.

Flame retardants, such as polybrominated diphenyl ethers (PBDEs), are equally harmful. These chemicals accumulate in the tissues of marine organisms, disrupting neurological function and reducing fertility. Research has found that PBDEs are present in significant concentrations in top marine predators, such as dolphins, seals, and large fish, due to biomagnification—the process by which contaminants become more concentrated as they move up the food chain. The neurological damage caused by PBDEs can impair the ability of these animals to hunt, navigate, and reproduce, leading to declines in their populations.

One well-documented example of the harmful effects of these chemicals is seen in killer whales (orcas). Orcas are among the most contaminated marine mammals, with high levels of PCBs, PBDEs, and other plastic-related pollutants found in their tissues. These chemicals have been linked to weakened immune systems and lower reproductive success, contributing to the decline of certain orca populations, particularly in areas with high levels of pollution, such as the Pacific Northwest.

The combined effects of microplastic ingestion and chemical contamination represent a significant and growing threat to marine biodiversity. As plastics continue to accumulate in the ocean, these pollutants are infiltrating ecosystems at all levels, from plankton to apex predators. Without intervention to reduce plastic waste and regulate the use of toxic chemicals in plastics, the damage to marine life will continue to escalate, with far-reaching consequences for ocean health and food security.

Nanoparticles, an emergent class of pollutants, pose a new and poorly understood threat to marine life. These ultrafine particles, typically smaller than 100 nm, are increasingly being released into the environment through a variety of industrial and consumer products. Common sources of nanoparticles entering the ocean include personal care products, industrial processes, and electronics manufacturing. For example, many sunscreens, cosmetics, and lotions contain nanoparticles of titanium dioxide (TiO_2) and zinc oxide (ZnO), which are used for their UV-blocking properties. These nanoparticles can be washed off the skin during swimming or showering, making their way into wastewater systems and eventually into the ocean.

In addition to personal care products, nanoparticles from industrial processes, such as paints, coatings, and fuel additives, also contribute to marine pollution. Electronics manufacturing is another significant source, as nanoparticles are used in the production of semiconductors and other high-tech materials. These tiny particles can enter the environment during production, use, and disposal of electronic devices, and once released into waterways, they eventually accumulate in the ocean. Even clothing made from nano-treated fabrics, which are designed to be stain resistant or antibacterial, releases nanoparticles into the water during washing cycles, further contributing to the ocean's nanoparticle load.

Nanoparticles have the potential to penetrate biological membranes due to their small size, potentially causing oxidative stress and damage at the cellular level in marine organisms. Research has shown that some nanoparticles can induce inflammation, disrupt metabolic processes, and even interfere with reproduction in marine species. For instance, studies have found that certain types of nanoparticles can accumulate in the tissues of fish, mollusks, and other marine animals, leading to physiological and developmental harm. Because nanoparticles can easily pass through the food chain, their effects could extend to larger predators, including humans who consume seafood.

The long-term impacts of nanoparticles on marine ecosystems are still being studied, but their widespread use and persistence in the environment raise significant concerns. Unlike larger particles of pollution, nanoparticles can interact with marine organisms at the cellular and molecular level, potentially causing harm that is difficult to detect and quantify. As the use of nanotechnology continues to grow, so too does the urgency for further research into the environmental and health consequences of these pollutants. Effective strategies to minimize nanoparticle pollution must be developed, focusing on better waste management, green chemistry innovations, and regulatory frameworks that address the unique risks posed by these ultrafine particles.

Pharmaceuticals, including medicines and commercial hormones, are also increasingly being detected in marine environments, often as a result of untreated or poorly treated wastewater entering rivers and oceans. These substances can have profound effects on marine species. Hormones such as synthetic estrogens, commonly found in birth control pills, can disrupt the endocrine systems of marine organisms, leading to reproductive abnormalities. For instance, exposure to these hormones has been linked to the feminization of male fish, where male individuals begin to develop female characteristics, such as producing eggs. This disruption of reproductive processes can lead to population declines in affected species. Similarly, other pharmaceuticals, such as antidepressants and antibiotics, can affect the behavior, growth, and immune response of marine life. Antidepressants, for example, have been shown to alter the behavior of fish, making them more vulnerable to predators. As these drugs accumulate in marine ecosystems, they pose a growing threat to biodiversity and ecosystem stability, highlighting the need for improved wastewater management to prevent further contamination.

Eutrophication, Dead Zones, and Ecosystem Collapse

Eutrophication is one of the most alarming consequences of nutrient pollution and is reshaping coastal and marine ecosystems around the globe. Caused primarily by the excessive input of nutrients like nitrogen and phosphorus from agricultural runoff, wastewater, and industrial discharges, eutrophication triggers the explosive growth of algae in marine waters. While algae are a natural part of ocean ecosystems, excessive blooms—often referred to as harmful algal blooms (HABs)—have catastrophic effects. These blooms deplete oxygen levels in the water, create hypoxic or "dead zones," and disrupt the balance of marine ecosystems.

One of the most destructive consequences of eutrophication in coastal systems is the formation of dead zones—areas of the ocean where oxygen levels are too low to support most forms of marine life. These zones occur when algal blooms die off and decompose, a process that consumes large amounts of dissolved oxygen in the water. As oxygen levels plummet, marine species either flee or die, leaving behind vast stretches of lifeless seabeds. The Gulf of Mexico, for example, experiences one of the largest dead zones in the world, stretching over 15,000 square kilometers (6334 square miles), largely due to nutrient runoff from the Mississippi River basin.

The Baltic Sea provides a stark example of how widespread and damaging eutrophication can be. With over 97,000 square kilometers affected by hypoxia and eutrophication, it is one of the most polluted seas in the world. The nutrient runoff from agriculture, combined with wastewater discharges from densely populated coastal regions, has led to massive algal blooms and extensive dead zones. These dead zones are so severe that large portions of the sea are nearly devoid of fish and other marine life. For commercial fisheries, the consequences have been dire, as key fish species like cod and herring struggle to survive in oxygen-deprived waters, threatening the livelihoods of communities that depend on these stocks.

Harmful algal blooms also produce toxins that directly impact marine life and human health. Several species responsible for red tides release neurotoxins that can kill fish, marine mammals, and seabirds. These toxins can also accumulate in shellfish, posing serious risks to humans who consume contaminated seafood, leading to illnesses such as paralytic and neurotoxic shellfish poisoning. In addition to the ecological damage, HABs inflict economic losses on coastal communities by impacting fisheries, tourism, and recreational activities. A notable example is the toxin produced by *Alexandrium* spp., which causes paralytic shellfish poisoning (PSP), a potentially fatal condition if contaminated shellfish are consumed. In the United States alone, PSP has resulted in numerous shellfish harvesting closures and, on occasion, human fatalities. Globally, thousands of cases of seafood poisoning from HABs are reported annually, with many going unreported in developing regions.

In addition to shellfish poisoning, HAB toxins can also cause respiratory problems when they become airborne. Along coastlines affected by red tides—blooms of toxic *Karenia brevis*, a dinoflagellate common in the Gulf of Mexico—waves and winds can release airborne toxins called brevetoxins. These can cause coughing, sneezing, and respiratory irritation in people near the shore, exacerbating conditions like asthma and affecting vulnerable populations, including the elderly and children. During the 2017–2018 red tide in Florida, reports of respiratory issues surged, with thousands of beachgoers experiencing symptoms due to airborne brevetoxins.

HABs also lead to substantial economic losses, particularly in coastal communities that rely on tourism, fishing, and shellfish harvesting. The 2006 HAB event in New England, which spanned from Maine to Massachusetts, led to an estimated $50 million in losses due to shellfish bed closures. The economic toll of HABs extends beyond direct impacts on fisheries and tourism; it includes the costs associated with public health interventions, lost jobs, and the long-term degradation of coastal ecosystems.

Globally, nutrient pollution is increasing, with an estimated 80% of marine pollution coming from land-based sources. The overuse of fertilizers in agriculture, untreated sewage, and industrial waste continues to drive eutrophication, leaving coastal waters vulnerable to dead zones and toxic blooms. Addressing this issue requires coordinated efforts to reduce nutrient inputs into marine environments, through better agricultural practices, stricter regulations on wastewater treatment, and improved land-use planning. Only by tackling the root causes of nutrient pollution can we hope to reverse the damage caused by eutrophication and prevent further collapse of marine ecosystems.

The cumulative effects of nutrient pollution, plastic waste, and chemical contaminants are leading to ecosystem collapse in many parts of the world's oceans. As key habitats such as coral reefs, mangroves, and seagrass beds degrade, marine biodiversity declines. The loss of these ecosystems not only impacts marine life but also reduces the ocean's ability to regulate carbon dioxide, protect coastlines from storm surges, and provide food for billions of people.

Human Health Impacts: Pollution in Our Food and Water

Marine pollution's impact on human health is becoming increasingly alarming, especially as more research uncovers the extent of contamination in seafood and other sources tied to the ocean. One of the most significant and well-documented threats comes from methylmercury, a toxic form of mercury that bioaccumulates in marine food chains. Predatory fish such as tuna, swordfish, and shark have been found to contain particularly high levels of methylmercury, making them a potential health hazard. The US Environmental Protection Agency has established a reference dose of 0.1 micrograms of methylmercury per kilogram of body weight per day, above which negative effects on human health, especially neurological damage, become more likely. However, as we previously stated, studies show that some fish contain methylmercury levels exceeding what is considered safe for human consumption over time. For pregnant women and young children, the risks are even greater, as methylmercury exposure can impair fetal brain development and lead to cognitive and motor deficits in children. In the United States alone, the Food and Drug Administration recommends limiting the consumption of high-mercury fish to once per week or less for these vulnerable groups.

The dangers of consuming seafood contaminated with toxic pollutants are not hypothetical but real and documented in several studies worldwide. For example, populations in Japan and other regions have historically suffered from mercury poisoning due to the consumption of contaminated seafood, with Minamata disease being a particularly tragic example of methylmercury poisoning. This disease, identified in the mid-twentieth century in Minamata, Japan, was caused by the release of methylmercury into the water by a chemical factory, which then bioaccumulated in fish and shellfish, leading to widespread poisoning. Victims of Minamata disease experienced severe neurological damage, including tremors, impaired vision, and speech, as well as developmental delays in children.

Beyond mercury, plastic pollution has emerged as a critical concern for human health. A 2019 study estimated that an adult in the United States may ingest up to 121,000 microplastic particles each year (the equivalent of a credit card per week) from various sources including air, tap water, and household dust, while seafood contributes only around 11,000 (less than 10%) particles annually for frequent consumers. Simply switching from tap water to bottled water can add up to 90,000 more microplastic particles per year. Moreover, daily activities such as washing synthetic clothing or heating food in plastic packaging often release far more microplastics than occasional seafood meals.

The health implications of microplastic ingestion and the chemicals associated to them are still being investigated, but initial studies suggest potential risks. Nevertheless, in reality, most of our exposure to microplastics stems from our everyday activities and immediate environment, rather than from consuming marine products. A study conducted by the World Health Organization in 2019 highlighted that while the health risks of microplastic ingestion remain uncertain, the presence of these particles in human food and water supplies is cause for concern. The study pointed out that microplastics, due to their small size, can penetrate human tissues and may lead to inflammation and oxidative stress, potentially causing long-term damage. Moreover, plastic additives can interfere with hormonal systems in humans, leading to reproductive issues, developmental problems, and increased risks of cancers such as breast and prostate cancer.

The Future of the Ocean and Human Health

The future health of the ocean and human populations are deeply intertwined, and the degradation of marine ecosystems presents significant risks that extend far beyond the environment itself. With over 3 billion people depending on the ocean for their primary source of protein, the collapse of marine ecosystems could trigger a cascade of socioeconomic and health crises. The ocean provides more than just food—it is a crucial player in regulating the climate, buffering against extreme weather events, and maintaining biodiversity, all of which directly or indirectly support human well-being. As pollution continues to push marine ecosystems toward collapse, the services they provide—such as food security, climate regulation, and disease control—are increasingly at risk.

The trajectory of marine degradation is particularly concerning when projections are made regarding future human populations. By 2050, the global population is expected to surpass 9 billion people, close to 10 billion, placing unprecedented pressure on food resources. The United Nations projects that the demand for seafood will continue to rise, with fisheries and aquaculture needing to supply nearly 200 million metric tons of seafood annually to meet global demand. Yet, many fish stocks are already overexploited, with some on the verge of collapse. If current trends in pollution, overfishing, and habitat destruction continue, many experts predict that the oceans could experience a collapse in fisheries by mid-century, with devastating consequences for global food security.

Climate change exacerbates these risks. Rising ocean temperatures and acidification are already altering marine ecosystems, leading to shifts in species distributions and declines in key fisheries. For example, the World Bank estimates that by 2050, climate change could reduce fish catches by as much as 30% in some regions, particularly in low-income countries that are heavily dependent on fisheries for both economic livelihoods and nutrition. This reduction in fish stocks is compounded by pollution, as the combination of toxic chemicals, plastics, and nutrient overloads degrades marine habitats, reduces biodiversity, and disrupts food chains.

In addition to food security, the health risks associated with marine pollution are expected to grow. As industrial activities continue to release pollutants into the ocean, the levels of hazardous substances like mercury, microplastics, and POPs are projected to rise. According to the United Nations Environment Programme, global plastic production is expected to double by 2040, which would drastically increase the amount of plastic waste entering the ocean. As plastics break down into microplastics and nanoplastics, their spread into marine food webs becomes more pervasive. Long-term exposure to these contaminants could lead to severe health impacts, especially in vulnerable populations such as pregnant women and children. Ingesting contaminated seafood with high levels of pollutants like methylmercury, plasticizers, and endocrine disruptors could lead to developmental issues, cognitive impairments, and increased risks of cancers.

Furthermore, climate change and pollution are likely to increase the incidence of HABs in the future. With coastal populations growing and tourism increasing in many regions, the risk of exposure to toxins derived by HABs, as well as the associated respiratory and gastrointestinal illnesses, will also grow. Projections suggest that the frequency and intensity of HABs could increase by 20–30% over the next few decades, posing further risks to both public health and coastal economies reliant on fisheries and tourism.

The World Health Organization and other global health bodies are already raising alarms about the potential for ocean-related health crises. As marine ecosystems continue to degrade, infectious diseases transmitted through waterborne pathogens are expected to rise. Coastal areas affected by pollution and nutrient overloads are becoming hot spots for bacterial growth, with pathogens such as *Vibrio* species—which thrive in warmer waters—becoming more prevalent. The risks of cholera outbreaks and other waterborne diseases could increase as water quality deteriorates, particularly in regions with inadequate sanitation and water treatment infrastructure.

Addressing these challenges will require urgent global cooperation, investment in sustainable practices, and the enforcement of stricter environmental regulations. The future of ocean health cannot be viewed in isolation from the well-being of human populations. Effective solutions will need to focus on mitigating pollution, protecting marine biodiversity, and building resilience against climate change. Without significant action, the degradation of marine ecosystems will continue to pose serious threats to human health, food security, and economic stability across the globe.

Further Reading

Clark, R. B. (2001). *Marine pollution* (5th ed., 237 p). Clarendon Press. A key textbook covering the range of marine pollutants, including classic pollutants like oil, nutrients, and heavy metals.

Cox, K. D., Covernton, G. A., Davies, H. L., Dower, J. F., Juanes, F., & Dudas, S. E. (2019). Human consumption of microplastics. *Environmental Science & Technology, 53*(12), 7068–7074.

Gallo, F., Fossi, C., Weber, R., Santillo, D., Sousa, J., Ingram, I., Nadal, A., & Romano, D. (2018). Marine litter plastics and microplastics and their toxic chemicals components: The need for urgent preventive measures. *Environmental Sciences Europe, 30*(1), 1–14. This article discusses both plastics and the toxic chemicals that adhere to marine debris, such as persistent organic pollutants (POPs).

Halpern, B. S., Walbridge, S., Selkoe, K. A., Kappel, C. V., Micheli, F., D'Agrosa, C., Bruno, J. F., et al. (2008). A global map of human impact on marine ecosystems. *Science, 319*(5865), 948–952. This article provides a comprehensive global map of the cumulative impacts of pollution and other human activities on marine ecosystems.

Jambeck, J. R., Geyer, R., Wilcox, C., Siegler, T. R., Perryman, M., Andrady, A., Narayan, R., & Law, K. L. (2015). Plastic waste inputs from land into the ocean. *Science, 347*(6223), 768–771. A foundational study estimating the amount of plastic entering the oceans annually, contextualizing the scale of the pollution crisis.

Kümmerer, K., Dionysiou, D. D., Olsson, O., & Fatta-Kassinos, D. (2018). A path to clean water. *Science, 361*(6399), 222–224. This article examines the environmental impacts of pharmaceuticals and personal care products (PPCPs) as emerging pollutants in water systems, including marine environments.

Richardson, S. D., & Ternes, T. A. (2018). Water analysis: Emerging contaminants and current issues. *Analytical Chemistry, 90*(1), 398–428. This review article covers emerging contaminants in marine environments, such as pharmaceuticals, personal care products, and endocrine-disrupting chemicals.

Rochman, C. M., Hoh, E., Hentschel, B. T., & Kaye, S. (2013). Long-term field measurement of sorption of organic contaminants to five types of plastic pellets: Implications for plastic marine debris. *Environmental Science & Technology, 47*(3), 1646–1654. Investigates how plastics in the ocean can absorb harmful chemicals, acting as vectors for pollutants like PCBs and flame retardants.

Weis, J. S. (2015). *Marine pollution: What everyone needs to know*. Oxford University Press. A readable guide discussing various forms of marine pollution, including emerging contaminants like pharmaceuticals and nanomaterials.

6

Biodiversity Under Threat

The ocean, home to the most diverse ecosystems on the planet, has been a cradle of life for billions of years. From sunlit seagrass meadows to the cold, dark abyss, the marine environment is teeming with organisms of all shapes and sizes. Marine biodiversity—ranging from tiny microorganisms like plankton to the great whales—plays an essential role in maintaining the health of our planet. Every species, no matter its size, is part of a complex and interconnected web that regulates critical Earth processes, including oxygen production, carbon sequestration, and climate stabilization. Yet, this biodiversity is now under unprecedented threat, primarily driven by human activities. The very survival of many marine species hangs in the balance, with consequences that will reverberate far beyond the ocean's boundaries.

The Accelerating Loss of Marine Life

While biodiversity is typically resilient to environmental changes over long geological time scales, the current rate of loss is unparalleled in human history. Anthropogenic pressures—climate change, overfishing, habitat destruction, pollution, and invasive species—are converging to create what scientists are increasingly referring to as the "sixth mass extinction." The difference between past extinctions and this one is that we are the primary driver. The accelerating rate of marine biodiversity loss is not only a concern for the distant future; it is happening now. Species are disappearing at an alarming pace, and entire ecosystems are on the brink of collapse.

Fig. 6.1 Example of coral suffering from bleaching. Philippines. (Author Albert Calbet)

Among the many human-induced pressures, climate change stands out as a particularly significant driver of biodiversity loss. Rising ocean temperatures, shifting currents, and ocean acidification are altering marine habitats at an unprecedented speed. Many species are unable to adapt quickly enough or migrate to more favorable environments. Tropical coral reefs, which support a quarter of all marine species, are particularly vulnerable to warming waters and increasingly frequent coral bleaching events.

Coral bleaching (Fig. 6.1) occurs when elevated water temperatures cause corals to expel the symbiotic algae that provide them with nutrients and their vibrant colors. Without these algae, corals turn white and, if the stress persists, die. The loss of coral reefs has cascading effects: countless species of fish, crustaceans, mollusks, and other organisms that rely on reefs for food, shelter, and breeding grounds are left vulnerable. Coral reefs are not just aesthetically stunning; they are crucial to marine biodiversity and the livelihoods of millions of people. Their decline represents a profound and far-reaching disruption to ocean ecosystems.

Fig. 6.2 Polar bear looking for food in the Arctic. (Author Albert Calbet)

Beyond coral reefs, polar regions—particularly the Arctic (Fig. 6.2)—are warming at more than twice the rate of the global average. The dramatic loss of sea ice is having destabilizing effects on species such as polar bears, walruses, and seals, which rely on ice for breeding, hunting, and resting. The disappearance of sea ice is altering entire ecosystems, beginning with the smallest organisms at the base of the food web. Phytoplankton and zooplankton, which thrive under the ice, are directly affected by the loss of habitat, leading to shifts in food availability for species further up the food chain, including fish, marine mammals, and seabirds. The Arctic is rapidly becoming a place where species that evolved for millennia are finding it harder to survive.

The open ocean is not exempt from the impacts of climate change. Warming waters and ocean acidification are altering marine food webs in ways that are less visible but equally destructive. Plankton are especially vulnerable to changes in ocean chemistry. Many species of plankton, such as foraminifera and coccolithophores, rely on calcium carbonate to build their shells. As the ocean becomes more acidic due to increased CO_2 absorption, these organisms struggle to form and maintain their shells, weakening the entire food chain. The collapse of plankton populations would have severe consequences for species that depend on them, from krill to whales, and would disrupt ecosystems across all oceanic regions.

Overfishing is one another of the most significant human pressures on marine biodiversity. Unsustainable fishing practices have depleted many fish stocks to the brink of collapse, with large predatory fish such as tuna, sharks, and swordfish seeing population declines of up to 90% in some regions. The loss of these apex predators has far-reaching consequences for marine ecosystems, as their absence allows smaller species to proliferate unchecked. This imbalance disrupts entire food webs, creating unpredictable changes in marine ecosystems.

For example, the decline of predatory fish often leads to the rise of opportunistic species such as jellyfish, which can outcompete fish for food and even prey on fish eggs and larvae. This phenomenon, known as trophic downgrading, shifts the balance of ocean ecosystems in ways that reduce biodiversity and diminish the ecological services that oceans provide. The consequences of overfishing extend beyond the environment; millions of people rely on fish for their primary source of protein and their livelihoods. The collapse of fish stocks threatens food security, particularly in developing nations where alternative sources of protein are limited.

Destructive fishing methods such as bottom trawling further exacerbate biodiversity loss. Bottom trawling involves dragging heavy nets along the seafloor, indiscriminately scooping up everything in their path, including non-target species and fragile ecosystems like coral reefs and seagrass meadows. These habitats, which act as nurseries for a wide variety of marine species, are often destroyed in the process, reducing biodiversity and weakening the resilience of marine ecosystems.

The devastation of marine habitats is another critical factor driving the loss of biodiversity. Coastal development, driven by urbanization, agriculture, and tourism, is encroaching on vital ecosystems such as mangroves, salt marshes, and estuaries. These coastal habitats are not just homes for marine species; they provide essential services, including acting as nurseries for fish, filtering pollutants, and protecting shorelines from erosion and storm surges. Their destruction compromises the ability of ecosystems to recover from environmental stresses and reduces the ocean's capacity to absorb CO_2, thus worsening climate change.

Particularly, sportive harbors can have significant negative effects on marine ecosystems. The construction process often involves dredging, land reclamation, and the installation of artificial structures, disrupting local habitats such as seagrass meadows, coral reefs, and sandy bottoms. The disruption of natural water flow can lead to stagnant areas that become ideal conditions for harmful algal blooms, further threatening biodiversity. Additionally, the change in circulation patterns can cause beach erosion, impacting coastal habitats and displacing the species that rely on them. Increased boat traffic also introduces pollution, noise, and physical disturbances, compounding the negative effects on marine ecosystems.

Large wind turbines are increasingly recognized as a vital component of the global transition to renewable energy, offering a sustainable solution to reduce carbon emissions and combat climate change. Their ability to generate significant amounts of clean electricity with minimal direct pollution makes them an essential part of the future energy mix. However, while the environmental benefits are clear, the ecological impact of wind farms, particularly offshore installations, has raised concerns. Wind turbines can disrupt local ecosystems, especially for bird and bat populations, which are vulnerable to collisions with turbine blades. While the underwater structures may provide new habitats for species that attach to surfaces, such as mussels and barnacles, they can also alter marine habitats, potentially affecting fish and invertebrate communities by changing water flow patterns, sediment deposition, and noise pollution. Moreover, the installation process itself, which involves pile driving and other activities, can temporarily disturb sensitive marine species, such as cetaceans and fish. Although wind energy is crucial for reducing reliance on fossil fuels, careful planning, monitoring, and mitigation strategies are needed to minimize its ecological footprint and protect biodiversity in both terrestrial and marine environments.

In addition to the destruction of coastal habitats, deep-sea environments are increasingly under threat from activities such as deep-sea mining and bottom trawling. These fragile ecosystems, many of which remain largely unexplored, are home to unique species that are poorly understood. Once destroyed, deep-sea habitats are slow to recover, if they recover at all, making the loss of biodiversity in these regions particularly alarming.

Invasive species, often introduced through global shipping, ballast water, and aquaculture, are further destabilizing marine ecosystems. Once introduced, invasive species can outcompete native species for resources, alter habitats, and disrupt food webs. For instance, the predatory lionfish, introduced into the Atlantic and Caribbean, has caused widespread damage to native fish populations, particularly on coral reefs. Invasive species are often highly adaptable and aggressive, making it difficult to control their spread once they are established.

The spread of invasive species is closely tied to globalization and the expansion of human activities. As global trade and travel increase, so does the likelihood of introducing species to new environments. The need for more stringent regulations and management of ballast water, aquaculture practices, and biosecurity is essential to prevent further introductions of invasive species.

Comparing Biodiversity Loss Across Land and Sea

When comparing biodiversity loss in marine ecosystems with that on land, both environments exhibit alarming declines, driven largely by human activities. However, the scope and mechanisms of biodiversity loss differ between the two realms, reflecting the unique challenges each faces. It is unquestionable that marine ecosystems are experiencing drastic biodiversity declines. Global fish stocks, for instance, have faced significant overexploitation, with 34% classified as overfished by 2017—an increase from 10% in 1974. Apex predators such as sharks and tuna have experienced severe declines due to both overfishing and habitat loss, with many species now critically endangered. Coral reefs face catastrophic losses, particularly under climate change scenarios. The Intergovernmental Panel on Climate Change (IPCC) warns that with a 1.5 °C increase in global temperature, 70–90% of coral reefs could disappear, and nearly all could vanish with a 2 °C rise. Furthermore, essential habitats like seagrass meadows and mangroves, which provide critical services such as coastal protection and carbon sequestration, have experienced declines of 29% and 20%, respectively, over the past century.

In terms of species loss, marine mammals have seen a 40% decline, and seabird populations have plummeted by 70% since 1970, according to the Living Planet Index. Iconic marine species such as the Steller's sea cow, the Caribbean monk seal, or the Japanese sea lion (and the freshwater Chinese river dolphin, Baiji) have already gone extinct, underscoring the urgent need for conservation efforts. Many other marine species, from large predators to microscopic organisms, are facing extinction pressures due to habitat degradation, overfishing, and pollution.

On land, biodiversity is also under immense pressure, with habitat loss being the primary driver of species extinction. According to the International Union for Conservation of Nature (IUCN), approximately 1 million species are at risk of extinction, with deforestation and land-use changes being the leading causes. The conversion of forests into agricultural land has particularly affected tropical rainforests, which harbor the most species-rich ecosystems on Earth. Since 1990, the world has lost approximately 420 million hectares of forest, significantly reducing habitat for terrestrial species.

Amphibians are particularly vulnerable, with around 40% of amphibian species at risk of extinction (Fig. 6.3), largely due to habitat destruction, pollution, and diseases such as chytridiomycosis. Bird populations have also faced steep declines, with one in eight bird species globally at risk of extinction due to habitat loss, hunting, and environmental changes. Mammals have seen

Fig. 6.3 Frogs waiting to be sold as food in a Hong Kong's Market. (Author Albert Calbet)

significant declines as well, with species such as the rhinoceros, tiger, and orangutan critically endangered due to poaching, deforestation, and fragmentation of their habitats.

Pollinators, essential for agricultural production, are in steep decline, with 40% of insect species—including bees, butterflies, and beetles—threatened with extinction. This decline has broader implications for global food security and ecosystem health. Terrestrial biodiversity loss is particularly acute in biodiversity hot spots such as Southeast Asia, the Amazon, and sub-Saharan Africa, where habitat conversion and deforestation rates are highest.

While both marine and terrestrial ecosystems face biodiversity crises, the drivers and manifestations of these declines differ. In the ocean, overfishing and climate change are the dominant pressures, with species loss driven largely by warming waters, ocean acidification, and habitat destruction. On land, habitat conversion (especially deforestation) is the primary driver of extinction, compounded by poaching, invasive species, and pollution.

The rate of species decline on land is comparable to that in the ocean, with both realms experiencing significant reductions in key species. Marine mammals have declined by 40%, a figure similar to the 40% extinction risk faced by amphibians on land. However, terrestrial ecosystems may face more acute challenges due to the fragmentation of habitats, which isolates species and reduces their ability to adapt to environmental changes. In summary, both marine and terrestrial biodiversity losses are driven by human actions, and the scale of the problem is similar, highlighting the urgent need for global action to preserve the planet's remaining biodiversity.

Restoring Marine Biodiversity: A Path Forward

Despite the grim outlook, there are efforts underway to mitigate biodiversity loss and restore marine ecosystems. The establishment of marine protected areas is one of the most effective tools for conserving marine biodiversity. Marine protected areas provide safe havens where human activities are restricted or prohibited, allowing ecosystems to recover and thrive. Studies show that well-managed marine protected areas can lead to significant increases in biodiversity, biomass, and ecosystem resilience. However, only a small fraction of the world's oceans is currently protected, and many marine protected areas lack adequate enforcement and resources.

Restoration projects are also showing promise. Coral reef restoration, for example, involves cultivating coral fragments in nurseries and transplanting them onto degraded reefs. Similarly, seagrass and mangrove restoration efforts are helping to rebuild critical habitats that have been lost to human activities. While these projects are often small in scale, they offer a glimpse of what is possible when science, policy, and community engagement come together to protect and restore marine ecosystems.

Addressing the crisis of biodiversity loss will require sustained global efforts across multiple sectors. The solutions are within reach, but time is running out. By acting decisively, we can protect the ocean's rich biodiversity and ensure the continued health of marine ecosystems for future generations.

Further Reading

Hughes, T. P., Kerry, J. T., Álvarez-Noriega, M., Álvarez-Romero, J. G., Anderson, K. D., Baird, A. H., et al. (2017). Global warming and recurrent mass bleaching of corals. *Nature, 543*(7645), 373–377. This study explores how global warming is altering the composition of coral reef ecosystems, particularly focusing on bleaching, with significant implications for marine biodiversity.

Jackson, J. B. C., Kirby, M. X., Berger, W. H., Bjorndal, K. A., Botsford, L. W., Bourque, B. J., Bradbury, R. H., et al. (2001). Historical overfishing and the recent collapse of coastal ecosystems. *Science, 293*(5530), 629–638. This classic paper documents the historical overfishing of marine ecosystems and its role in the collapse of coastal biodiversity.

McCauley, D. J., Pinsky, M. L., Palumbi, S. R., Estes, J. A., Joyce, F. H., & Warner, R. R. (2015). Marine defaunation: Animal loss in the global ocean. *Science, 347*(6219), 1255641. This paper discusses the loss of large animals from marine ecosystems (marine defaunation) and its consequences for ocean biodiversity.

Mora, C., Tittensor, D. P., Adl, S., Simpson, A. G., & Worm, B. (2011). How many species are there on Earth and in the ocean? *PLoS Biology, 9*(8), e1001127. An influential paper that estimates the number of species on Earth, including marine species, and the challenges in conserving marine biodiversity.

Pauly, D., & Zeller, D. (Eds.). (2016). *Global atlas of marine fisheries: A critical appraisal of catches and ecosystem impacts* (520 p). Island Press. This book provides a detailed examination of global fisheries, highlighting the consequences of overfishing and the decline in marine biodiversity.

Pimm, S. L., Jenkins, C. N., Abell, R., Brooks, T. M., Gittleman, J. L., Joppa, L. N., et al. (2014). The biodiversity of species and their rates of extinction, distribution, and protection. *Science, 344*(6187), 1246752 This paper provides an assessment of global biodiversity, extinction rates, and the need for increased protection, including a focus on marine species.

Roberts, C. (2012). *The ocean of life: The fate of man and the sea* (405 p). Penguin Books. A comprehensive look at the state of the world's oceans and the human impact on marine biodiversity, with a focus on overfishing, pollution, and climate change.

Sala, E., & Knowlton, N. (2006). Global marine biodiversity trends. *Annual Review of Environment and Resources, 31*(1), 93–122. Provides a comprehensive review of global trends in marine biodiversity, highlighting the main drivers of biodiversity loss.

7

The Future of Marine Plankton

Marine plankton, though often overlooked in climate change discussions, are fundamental to life on Earth. These tiny organisms not only form the foundation of ocean ecosystems but also regulate global climate systems and offer potential solutions to feeding a growing global population. As the planet faces unprecedented environmental shifts, the fate of plankton becomes a central issue for both ecological and human survival. This chapter delves into the projected future of plankton, exploring their critical role in climate regulation, ecosystem balance, and their emerging importance in addressing global food security.

Plankton and Climate Regulation: Earth's Hidden Thermostat

Plankton, particularly phytoplankton (Fig. 7.1), play a pivotal role in regulating Earth's climate by acting as one of the planet's largest carbon sinks. The ocean's role as a carbon sink relies on two primary mechanisms: the physical carbon pump and the biological carbon pump. The physical pump refers to the ocean's ability to absorb CO_2 at the surface, particularly in cold waters, where the gas dissolves more easily. Ocean currents then carry this carbon-rich water to the deep ocean, where it is stored for centuries or longer. The biological pump works through marine organisms like phytoplankton, which absorb CO_2 through photosynthesis. When they die or are consumed by zooplankton, a portion of the carbon they contain sinks as marine snow to the ocean

Fig. 7.1 Phytoplankton. Diatom chain. (Author Albert Calbet)

floor, sequestering it in deep-sea sediments for centuries. Together, these processes have kept Earth's climate in balance for millions of years.

It is estimated that phytoplankton sequester 2–3 billion tons of carbon annually, significantly reducing atmospheric CO_2. Without this process, the Earth's atmosphere would contain approximately 50% more CO_2 than it does today, drastically increasing global temperatures. However, as ocean temperatures rise and stratification increases, the efficiency of this pump is projected to decline. By 2100, some models predict a significant reduction in phytoplankton biomass, particularly in tropical and subtropical regions, where nutrient availability will diminish due to decreased vertical mixing.

This reduction could have dire consequences for global carbon sequestration. As the productivity of phytoplankton declines, less CO_2 will be absorbed, accelerating climate change. This feedback loop poses a grave risk, as a failure in the ocean's carbon pump could exacerbate the already critical warming trends, leading to runaway climate effects. In this scenario, plankton will be unable to sustain their role as Earth's climate regulators, setting the stage for more extreme and unpredictable weather patterns, rising sea levels, and habitat destruction.

Plankton Shifts: Ecosystem Mismatches and Phenological Crises

In response to warming waters, both phytoplankton and zooplankton (Fig. 7.2) are migrating toward the poles. This poleward migration alters the composition of species in marine food webs, introducing new organisms that

Fig. 7.2 Zooplankton. Copepod. (Author Albert Calbet)

are not always suitable for the existing ecosystems. Additionally, warming waters can cause phenological mismatches, where the timing of plankton blooms no longer aligns with the life cycles of species that depend on them. For example, fish larvae in regions such as the North Atlantic rely on synchronized blooms of zooplankton for nourishment. If these blooms occur earlier or later than expected, fish populations may be left without sufficient food during critical stages of development, leading to potential fisheries collapses.

By mid-century, entire marine ecosystems may be reshaped as plankton species abandon equatorial waters and temperate zones for cooler, more habitable regions. The Arctic and Antarctic could become hot spots for plankton activity, but the influx of temperate species will place immense pressure on native polar ecosystems, potentially driving out key species like krill, which form the base of polar food webs. Such shifts could result in cascading effects throughout the ocean, with large predators such as whales, penguins, and seals facing starvation as their primary food sources disappear or decline.

Compounding these challenges, the decline of apex predators—such as sharks, tuna, and large marine mammals—due to overfishing and habitat loss can further destabilize plankton populations. Apex predators play a key role in maintaining the balance of marine ecosystems by regulating the populations of mid-level predators and herbivores. As apex predator populations diminish, mid-level species such as small fish and jellyfish may experience population booms, which can result in increased grazing on zooplankton. This imbalance can lead to a decline in zooplankton populations, further affecting the food web. In addition, some mid-level predators that thrive in the absence of apex species may consume large amounts of phytoplankton or graze on smaller zooplankton, disrupting the natural carbon cycle and reducing the efficiency of the biological carbon pump. Ultimately, the loss of apex predators can trigger cascading effects throughout the ecosystem, exacerbating the already precarious situation for plankton and the marine food web as a whole.

Moreover, rising temperatures and decreasing nutrient availability are expected to shift the size structure of phytoplankton communities. Warmer oceans, combined with increased stratification, limit the vertical mixing of nutrient-rich deeper waters to the surface, where phytoplankton thrive. Under such nutrient-poor conditions, smaller phytoplankton species—such as picoplankton and nanoplankton—are likely to become more dominant. These smaller phytoplankton species have higher surface-area-to-volume ratios, making them better adapted to low-nutrient environments. However, smaller phytoplankton are less efficient at sequestering carbon compared to their larger counterparts, such as diatoms, which play a crucial role in the biological carbon pump by sinking quickly to the ocean floor after they die.

The dominance of smaller phytoplankton species could reduce the overall carbon sequestration potential of the ocean, weakening one of the Earth's primary mechanisms for regulating atmospheric CO_2. Additionally, smaller phytoplankton support different food webs, often dominated by smaller zooplankton, jellyfish, and other gelatinous plankton, rather than larger species like krill, which are more important for sustaining higher predators like fish, whales, and seabirds. This shift in phytoplankton size could further alter marine ecosystems, reducing the food availability for larger species and ultimately threatening biodiversity, fisheries, and the stability of marine ecosystems.

Expanding Oxygen Minimum Zones

Oxygen minimum zones are already a significant feature in certain parts of the ocean, especially in the Eastern Tropical Pacific, the Arabian Sea, and parts of the Bay of Bengal. Currently, these zones occupy around 8% of the global ocean. Projections suggest that by 2100, oxygen minimum zones could expand by up to 50% (an additional 4% of the ocean) due to ongoing warming, especially in the Pacific and Indian Oceans.

While eutrophication has already been stressed as responsible of oxygen minimum zones, a key driver of their expansion is the increased stratification of the ocean caused by surface warming. As surface waters warm, they become lighter and less likely to mix with deeper, cooler, nutrient-rich waters. This reduced vertical mixing limits the amount of oxygen that can be transferred to deeper layers, leading to the intensification and spread of oxygen minimum zones. Research indicates that the oxygen content of the global ocean has already decreased by approximately 2% since the mid-twentieth century, a trend that is expected to continue as global temperatures rise.

Plankton, particularly zooplankton, are highly sensitive to oxygen levels. As oxygen minimum zones expand, these organisms will be forced into shrinking oxygen-rich habitats, a phenomenon known as habitat compression. In the Eastern Tropical Pacific, for example, large zooplankton such as copepods and krill are already experiencing the effects of shrinking oxygen availability. As they are confined to shallower, oxygen-rich zones, competition for food and space increases, leading to reduced biodiversity. Studies show that plankton biomass in oxygen minimum zones can be up to 30% lower than in well-oxygenated waters, highlighting the severe impact of expanding oxygen minimum zones on plankton communities.

The expansion of oxygen minimum zones will also alter the species composition of plankton communities. Jellyfish and other gelatinous organisms are more tolerant of low oxygen conditions and may become dominant in these regions. For instance, in the Eastern Tropical North Pacific, jellyfish populations have been observed to increase in oxygen minimum zones, while more diverse and productive species like krill and copepods decline. This shift in community structure can destabilize marine ecosystems, as jellyfish are inefficient at transferring energy through the food web compared to other zooplankton species. The increased presence of gelatinous organisms could lead to food shortages for higher predators, such as fish, sharks, and whales, which depend on more nutrient-rich plankton species to sustain their populations.

The loss of diverse and productive plankton species will weaken entire marine food chains. For example, in the Arabian Sea, where oxygen minimum zones have been expanding, fish biomass has decreased, affecting fisheries that are economically important to coastal communities. Globally, as oxygen minimum zones grow, the reduced plankton diversity will diminish the resilience of marine ecosystems, making them more vulnerable to environmental shocks such as temperature fluctuations, acidification, and overfishing. The expansion of oxygen minimum zones presents a significant threat not only to plankton communities but to the broader health of ocean ecosystems and the species that depend on them.

Plankton as a Food Source for an Overpopulated Planet

As global population growth accelerates, with projections suggesting a peak of nearly 10 billion people by 2050, food security becomes a critical issue. Conventional agriculture is already straining under the pressure of feeding a growing population, and climate change is exacerbating the challenges by reducing crop yields and increasing the frequency of extreme weather events. In this context, plankton—particularly phytoplankton—offer a promising solution to addressing future food shortages.

Phytoplankton, especially microalgae such as spirulina and chlorella, are highly nutritious, containing essential fatty acids, proteins, vitamins, and antioxidants. They are already being cultivated for use in nutritional supplements, animal feed, and biofuels, but their potential as a direct food source for humans remains largely untapped. Phytoplankton farming could provide a sustainable and low-impact alternative to traditional agriculture. These organisms grow rapidly, require minimal freshwater, and can be cultivated in coastal areas or controlled bioreactors, reducing the environmental footprint of food production.

In the future, large-scale plankton aquaculture could offer a new source of nutrition that requires fewer resources than traditional farming. Unlike crops, which are vulnerable to drought, pests, and land degradation, phytoplankton can be grown in controlled environments, where their growth can be optimized to ensure consistent yields. By incorporating phytoplankton into global food systems, humanity could address critical issues related to food security while reducing the environmental impacts of agriculture, such as deforestation, freshwater depletion, and greenhouse gas emissions.

Additionally, zooplankton such as krill and copepods offer potential as a food source. They are already harvested for their oil, which is rich in omega-3 fatty acids, but their use as a direct food source could expand in the coming decades. High in protein and essential nutrients, krill and copepods could be integrated into fish meal and human diets as a sustainable protein alternative. However, krill fishing must be carefully managed to avoid further destabilizing marine ecosystems, particularly in polar regions where krill are a keystone species.

Challenges and Risks of Plankton-Based Food Production

Despite the promising potential of plankton as a future food source, there are significant challenges to overcome. The large-scale farming of plankton, particularly in ocean-based systems, could exacerbate existing environmental pressures. Overharvesting phytoplankton or zooplankton from the wild could destabilize marine food webs, reducing the availability of essential nutrients for species like fish and whales. Additionally, plankton farming could contribute to ocean acidification, nutrient depletion, and pollution if not managed carefully.

Innovative solutions, such as plankton farming, offer a way to decouple plankton production from natural ecosystems. By growing plankton in closed systems or bioreactors, scientists could optimize yields without disrupting marine habitats. This approach also reduces the risk of introducing invasive species into new regions, a concern with ocean-based farming. However, the technological and financial barriers to large-scale plankton production must be addressed to make these systems viable on a global scale.

Another challenge is consumer acceptance. While microalgae-based products like spirulina and chlorella are gaining popularity, widespread adoption of plankton as a staple food may face cultural and economic hurdles. Educating the public about the nutritional and environmental benefits of plankton-based foods will be critical in overcoming resistance and integrating these organisms into mainstream diets.

In an era where climate change, population growth, and resource scarcity are converging into a global crisis, plankton could become a vital part of the solution. These tiny organisms not only regulate the Earth's climate and support marine ecosystems but also offer a sustainable, nutrient-rich alternative to traditional agriculture. As food insecurity looms for future generations, investing in plankton-based aquaculture could help ensure a stable and resilient global food supply.

By protecting and responsibly managing plankton populations, humanity can harness their potential to support a growing population while safeguarding the health of the planet. Whether through the development of synthetic plankton farms or the expansion of sustainable phytoplankton cultivation, these microscopic organisms could become the foundation of future food systems—providing not only a solution to global hunger but also a way to mitigate the environmental impacts of conventional agriculture.

Further Reading

Behrenfeld, M. J., & Boss, E. (2018). Student's tutorial on bloom hypotheses in the context of phytoplankton annual cycles. *Global Change Biology, 24*(1), 55–77. This review explores phytoplankton bloom dynamics and how environmental changes, such as temperature shifts and nutrient availability, can alter the timing and extent of these blooms.

Boyd, P. W., & Doney, S. C. (2002). Modelling regional responses by marine pelagic ecosystems to global climate change. *Geophysical Research Letters, 29*(16), 1806. This study uses models to predict how marine ecosystems, especially plankton, will respond to climate change. It highlights the importance of plankton in global carbon cycling and how ocean warming and acidification could disrupt these processes.

Calbet, A. (2024). *Plankton in a changing world: The impact of global change on marine ecosystems.* Springer Nature. This forthcoming book examines how global change is impacting plankton and marine ecosystems, addressing the role of plankton in global biogeochemical cycles.

Calbet, A., & Landry, M. R. (2004). Phytoplankton growth, microzooplankton grazing, and carbon cycling in marine ecosystems. *Limnology and Oceanography, 49*(1), 51–57. This research examines the interplay between phytoplankton growth and zooplankton grazing, illustrating how these interactions influence carbon cycling and ecosystem structure in marine environments.

Falkowski, P. G. (2012). The power of plankton. *Nature, 483*, S17–S20. This paper explores the role of plankton in regulating Earth's climate and carbon cycle, emphasizing how plankton, particularly phytoplankton, sequester carbon, impacting both the marine ecosystem and global climate systems.

Hutchins, D. A., & Fu, F. X. (2017). Microorganisms and ocean global change. *Nature Microbiology, 2*, 17058. This article examines how climate change affects marine microorganisms, including phytoplankton, and discusses potential consequences for marine ecosystems and biogeochemical cycles.

Moore, J. K., et al. (2018). Sustained climate warming drives declining marine biological productivity. *Science, 359*(6380), 1139–1143. This study projects a long-term decline in marine biological productivity, including phytoplankton, due to sustained ocean warming and stratification, with serious consequences for marine biodiversity and global carbon cycles.

Sommer, U., & Lewandowska, A. (2011). Climate change and the timing, magnitude, and composition of the phytoplankton spring bloom. *Global Change Biology, 17*(2), 791–803. The paper investigates how climate change affects the timing and composition of phytoplankton blooms, which could have profound implications for marine food webs and carbon sequestration.

Steinberg, D. K., & Landry, M. R. (2017). Zooplankton and the ocean carbon cycle. *Annual Review of Marine Science, 9*, 413–444. This review highlights the critical role of zooplankton in the biological carbon pump, examining how zooplankton dynamics influence carbon fluxes in marine ecosystems.

8

The Rise of Jellyfish Dominance

As the ocean faces unprecedented environmental pressures, a growing concern among marine scientists is the increasing dominance of jellyfish. In many regions, jellyfish populations are proliferating at alarming rates, disrupting ecosystems, fisheries, and even coastal economies. This trend raises a critical question: could the ocean of tomorrow be one where jellyfish, rather than fish, are the dominant species? Although this may seem like an exaggerated or unlikely scenario, mounting evidence suggests that jellyfish are thriving in the very conditions humans have created, indicating a fundamental shift in marine ecosystems.

Resilience of an Ancient Organism

Jellyfish (Fig. 8.1) are indeed among the oldest multicellular organisms on Earth, with a lineage stretching back over 500 million years. Their long evolutionary history has endowed them with a remarkable ability to adapt to changing environmental conditions, allowing them to survive multiple mass extinctions and shifts in global ecosystems. This resilience is largely due to their simple body structure, versatile reproductive strategies, and ability to thrive in a wide range of environmental conditions. Jellyfish are primarily composed of water and are capable of surviving in nutrient-poor environments where more complex organisms would struggle. This simplicity is a key factor in their ability to endure dramatic changes in the Earth's oceans over millennia.

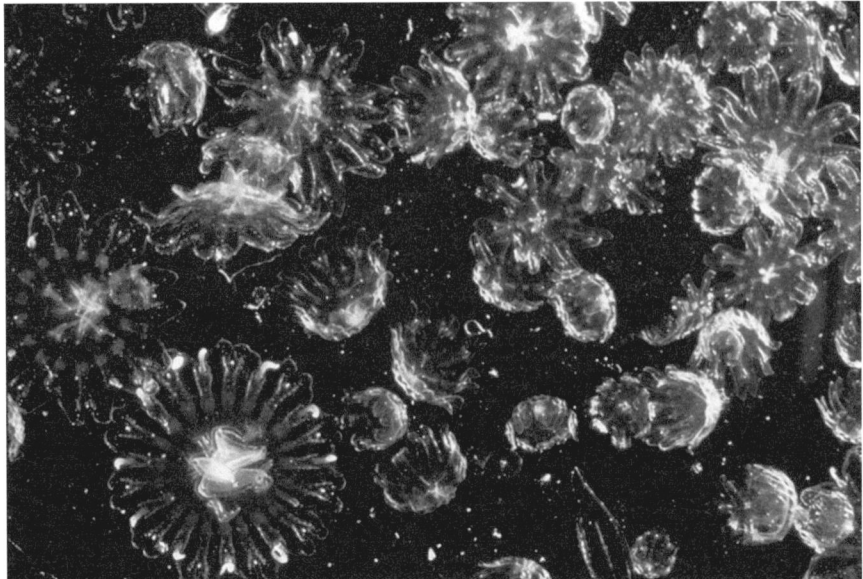

Fig. 8.1 Ephyra of jellyfish. (Author. C. Carré)

However, the recent global surge in jellyfish populations is not just a reflection of their ancient resilience but is closely tied to human-induced changes in the marine environment. Overfishing, climate change, pollution, and habitat degradation have collectively created conditions that allow jellyfish populations to thrive while other marine species struggle. In many cases, human activities have removed key predators and competitors, giving jellyfish the opportunity to proliferate unchecked.

The Impact of Overfishing: Declining Predators

Overfishing has dramatically reshaped marine ecosystems by removing large predators that traditionally keep jellyfish populations in check. Species such as tuna, swordfish, sharks, and sea turtles—natural jellyfish predators—have been heavily overfished in many regions, resulting in a significant decline in predation pressure on jellyfish. A report by the Food and Agriculture Organization (FAO) estimates that around 34% of global fish stocks are overfished, with many predator species facing population declines of up to 90% in certain areas. For instance, large predatory fish like bluefin tuna, which feed on jellyfish in their juvenile stages, have seen drastic population decreases in the Mediterranean due to unsustainable fishing practices. This creates an ecological vacuum, allowing jellyfish populations to grow unchecked.

In the Black Sea, the overfishing of small pelagic fish, such as anchovies and sprats, helped the rise of the invasive comb jelly (*Mnemiopsis leidyi*). This species proliferated in the absence of predators, consuming large amounts of zooplankton and collapsing local fish stocks. Similar scenarios have played out in other regions, with overfishing exacerbating jellyfish blooms in areas like the East China Sea, Gulf of Mexico, and Baltic Sea.

Overfishing does not just remove the direct predators of jellyfish but also disrupts the entire food web. When predatory fish are overexploited, smaller fish and plankton-eating species thrive, which may lead to increased competition for food with jellyfish. Additionally, some of these small species may feed on the same zooplankton that jellyfish consume, indirectly intensifying competition and potentially triggering trophic cascades that further benefit jellyfish proliferation.

Climate Change: Warming Waters and Expanding Habitats

Climate change is another significant factor driving the global increase in jellyfish populations. Rising sea temperatures, caused by anthropogenic climate change, have expanded the habitable zones for jellyfish, allowing them to thrive in areas where they were previously limited. According to the Intergovernmental Panel on Climate Change (IPCC), the world's oceans have warmed by about 0.88 °C on average since 1850, with some regions, such as the Mediterranean Sea, experiencing even faster warming.

Jellyfish are particularly well suited to warmer waters due to their ability to tolerate a broad range of temperatures. Studies have shown that warmer seas increase jellyfish reproduction rates, leading to more frequent and larger blooms. In the Mediterranean, where water temperatures are rising faster than the global average, the frequency of jellyfish blooms has increased significantly in recent decades. For example, *Pelagia noctiluca*, a common jellyfish species in the region, has experienced population booms, with blooms observed nearly every summer since 2000, compared to every 12 years in the 1980s. These blooms have had serious economic consequences for local fisheries and tourism, with some Mediterranean countries spending millions of euros annually to manage jellyfish outbreaks and protect beachgoers.

In addition to temperature, climate change is altering ocean circulation patterns, creating favorable conditions for jellyfish. As surface waters warm, the oceans are becoming more stratified, with less mixing between the nutrient-rich deep waters and the surface. This stratification reduces nutrient availability for phytoplankton, the base of the marine food web, creating conditions where jellyfish—known for their ability to survive in nutrient-poor waters—thrive. Regions like the North Pacific and South Atlantic have already observed shifts in jellyfish species moving poleward as warmer waters expand.

Pollution: Fueling Jellyfish Blooms

Pollution, particularly nutrient pollution from agricultural runoff and untreated sewage, is another key factor driving jellyfish blooms. Excess nutrients such as nitrogen and phosphorus from fertilizers enter coastal waters, fueling algal blooms in a process known as eutrophication. When these algal blooms die and decompose, they deplete oxygen levels in the water, creating hypoxic zones, where most marine life struggles to survive. Jellyfish, however, are highly tolerant of low-oxygen conditions and can outcompete other species in these environments.

In the hypoxic conditions of the Gulf of Mexico, jellyfish species such as *Aurelia aurita* (Fig. 8.2) and *Chrysaora quinquecirrha* have flourished, taking advantage of reduced competition from oxygen-dependent species. In the Baltic Sea, another major dead zone, jellyfish populations have increased in parallel with declining oxygen levels. The combination of eutrophication and overfishing has led to jellyfish dominance in some areas, displacing native species and disrupting fisheries. This shift toward jellyfish-dominated ecosystems is particularly concerning for the sustainability of the region's commercial fish stocks, such as cod and herring, which are already under pressure from overfishing.

Impact on Economy and Infrastructures

Jellyfish blooms are becoming an increasing concern not only for marine ecosystems but also for the global economy, especially in coastal regions where tourism and infrastructure are vital. Coastal tourism, a multibillion-dollar industry, is particularly vulnerable to jellyfish population booms. In popular tourist destinations, such as the Mediterranean, Japan, and Australia, the rise in jellyfish stings has become a serious issue. Beachgoers, unaware of the risks,

Fig. 8.2 Jellyfish. (Author Albert Calbet)

are often surprised by jellyfish swarms, leading to painful stings, some of which can be fatal, depending on the species. For example, the box jellyfish (*Chironex fleckeri*), found in Australia and Southeast Asia, is among the most venomous creatures on Earth, with its sting capable of causing severe pain, paralysis, and even death within minutes. As these incidents increase, local authorities are forced to close beaches, leading to a decline in tourism and economic loss for coastal businesses.

The Mediterranean, one of the most heavily trafficked tourist destinations in the world, has seen a notable rise in jellyfish stings. In Spain alone, thousands of beach closures and advisories occur each summer due to jellyfish. For example, in 2018, large swarms of the *Pelagia noctiluca* jellyfish led to the closure of several beaches along Spain's Costa del Sol, impacting local businesses dependent on tourist traffic. In Italy, stings by *Pelagia noctiluca* have also surged, with over 40,000 recorded jellyfish stings along the Adriatic coastline in 2013. The economic cost of these blooms, in terms of lost revenue from tourism and the cost of beach management and medical treatments, is difficult to quantify but is estimated to be in the millions of euros annually for each affected region.

Aside from the direct impact on human health and tourism, jellyfish are also disrupting critical infrastructure. In numerous instances, jellyfish blooms have clogged the intake pipes of power plants, desalination facilities, and industrial ships. For example, in 2011, a jellyfish bloom caused the shutdown of a nuclear power plant in Scotland after jellyfish blocked its cooling water intake. A similar incident occurred in Sweden, where the Oskarshamn nuclear power plant was forced to temporarily shut down in 2013 after a large swarm of moon jellyfish blocked the plant's cooling system. Desalination plants, which are critical for supplying freshwater in arid regions, have also faced operational challenges due to jellyfish clogging their intake systems, particularly in countries such as Israel and Saudi Arabia, where desalination plays a crucial role in water security. The costs of these shutdowns include not only lost production time but also the expense of clearing out the jellyfish and repairing damaged equipment, which can run into the millions of dollars for a single incident.

Shipping lanes and fisheries are similarly impacted by the rise of jellyfish populations. Jellyfish dominance creates what scientists call "trophic dead ends." Unlike fish, jellyfish (Fig. 8.3) do not efficiently transfer energy up the food chain. In ecosystems dominated by jellyfish, energy remains trapped in jellyfish populations rather than being passed on to higher trophic levels, such as fish, seabirds, and marine mammals. This leads to reduced productivity and biodiversity in affected ecosystems, further weakening the resilience of marine environments. Moreover, large jellyfish swarms can clog fishing nets, damage boats, and disrupt fishing operations. In Japan, the massive Nomura jellyfish (*Nemopilema nomurai*), which can reach up to two meters in diameter and weigh over 200 kilograms, has caused severe disruptions to the country's fishing industry. In 2009, a fleet of Japanese fishing boats capsized when they tried to haul in their nets, which were weighed down by thousands of Nomura jellyfish. These jellyfish not only disrupt fishing activities but also destroy fish stocks by feeding on fish eggs and larvae, further depleting marine resources. In South Korea, the fishing industry has suffered similar economic losses, with jellyfish blooms causing a significant reduction in fish catches, estimated to cost the country's fishing industry over $100 million annually.

8 The Rise of Jellyfish Dominance

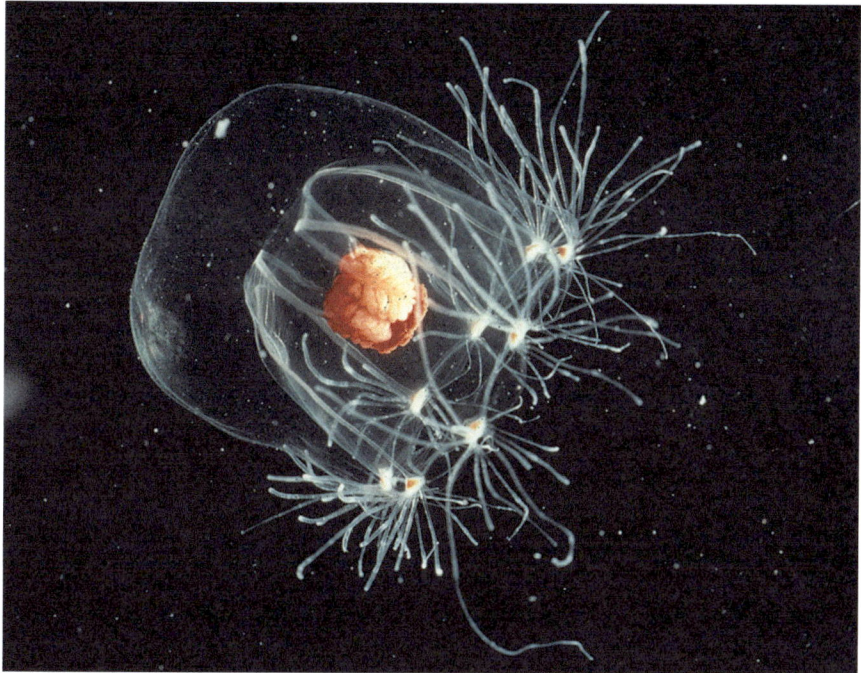

Fig. 8.3 Jellyfish. (Author C. Carré)

The increasing dominance of jellyfish in marine ecosystems is a stark indicator of environmental degradation. Overfishing, climate change, pollution, and habitat destruction have contributed to the collapse of fish populations and the decline of marine predators that typically control jellyfish numbers, such as sea turtles and large predatory fish like tuna and sharks. This ecological imbalance has allowed jellyfish, opportunistic and resilient species, to proliferate. Warmer seas, a by-product of climate change, have created ideal conditions for jellyfish, as these organisms are highly adaptable to temperature fluctuations. Additionally, jellyfish thrive in low-oxygen environments such as "dead zones" created by nutrient pollution from agricultural runoff and untreated sewage. In these hypoxic conditions, jellyfish often outcompete fish and other marine species, further contributing to the degradation of marine ecosystems.

The rise of jellyfish also represents a significant threat to biodiversity. In ecosystems dominated by jellyfish, energy does not flow efficiently up the food chain, leading to trophic "dead ends." Jellyfish consume vast amounts of plankton and small fish, yet they are rarely consumed by higher predators. This means that energy remains trapped within jellyfish populations, rather than being transferred to higher trophic levels, such as fish, seabirds, or marine mammals. The result is less productive ecosystems that are unable to support the same levels of biodiversity and biomass as they once did. Coral reefs, for example, are particularly vulnerable to jellyfish proliferation. In regions where overfishing and warming waters have decimated fish populations, jellyfish blooms are putting additional stress on coral ecosystems, which are already struggling to cope with coral bleaching and ocean acidification.

In countries like China, jellyfish are harvested as a delicacy, and there is growing interest in expanding the market for edible jellyfish. Some researchers are exploring the potential of jellyfish as a sustainable seafood alternative, particularly in the face of declining fish stocks. In China and Japan, some species are processed and consumed in various dishes. However, not all jellyfish are edible, and the sheer scale of some blooms, like those of the Nomura jellyfish, creates problems that far outweigh the economic benefits of harvesting them. In many cases, jellyfish blooms are so large that they overwhelm local ecosystems and economies, causing significant disruption to fishing operations, coastal infrastructure, and tourism.

Jellyfish blooms serve as a reminder of the fragility of marine ecosystems and the importance of maintaining balance in the ocean. The factors driving the rise of jellyfish—overfishing, climate change, pollution, and habitat loss—are all products of human activity. The increasing dominance of jellyfish in many regions is a symptom of the broader environmental crisis facing the world's oceans. If current trends continue, we may see entire sections of the ocean dominated by jellyfish, with far-reaching consequences for biodiversity, fisheries, and human economies. The future of marine ecosystems may depend on addressing the root causes of jellyfish proliferation and restoring balance to these vital environments.

Preventing a Jellyfish-Dominated Future

To prevent jellyfish from becoming the dominant species in marine ecosystems, it is crucial to recognize that jellyfish are not the enemy but a mere consequence of human actions that have disrupted the natural balance of the ocean. Instead of focusing efforts on removing or fishing out jellyfish, we must address the underlying causes that have allowed their populations to explode and focus on creating conditions that support the recovery of other marine species. Sustainable fishing practices are essential to rebuilding populations of large predatory fish, such as tuna, swordfish, sharks, and sea turtles, which naturally help regulate jellyfish numbers. Overfishing has diminished these species, creating an imbalance that jellyfish can exploit. Pollution, particularly nutrient runoff from agriculture and untreated sewage, leads to eutrophication and the creation of hypoxic dead zones where jellyfish thrive. Tackling this problem requires better waste management practices and reducing the use of fertilizers, alongside stronger regulations to limit plastic pollution, which further disrupts marine ecosystems. Another significant factor is climate change, which is warming the oceans and making them more conducive to jellyfish proliferation. Reducing greenhouse gas emissions is critical to slowing ocean warming and helping ecosystems remain resilient. Global climate action, such as the Paris Agreement, must be strengthened to mitigate the worst impacts on marine life. Additionally, protecting and restoring key marine habitats, such as coral reefs, seagrass meadows, and mangroves, can provide the necessary support for fish populations to thrive and outcompete jellyfish. Expanding and enforcing marine protected areas can help restore biodiversity and foster healthier ecosystems that limit jellyfish blooms. Finally, raising public awareness about the root causes of jellyfish dominance is essential. People must understand that jellyfish blooms are a symptom of human activities, and the solution lies in addressing these broader issues to restore the health of marine ecosystems. By creating better conditions for diverse marine life to recover, we can naturally control jellyfish populations and restore balance to the oceans.

Further Reading

Condon, R. H., Duarte, C. M., Pitt, K. A., et al. (2013). Recurrent jellyfish blooms are a consequence of global oscillations. *Proceedings of the National Academy of Sciences, 110*(3), 1000–1005. This research suggests that jellyfish blooms are influenced by global climatic patterns, particularly large-scale oscillations like the North Atlantic Oscillation, which drive environmental changes conducive to jellyfish population growth.

Dong, Z., Liu, D., & Keesing, J. K. (2010). Jellyfish blooms in China: Dominant species, causes, and consequences. *Marine Pollution Bulletin, 60*(7), 954–963. This study examines the increasing frequency of jellyfish blooms along China's coasts, analyzing the environmental and anthropogenic factors contributing to the blooms and their impacts on fisheries and coastal economies.

Duarte, C. M., Pitt, K. A., Lucas, C. H., et al. (2013). Is global ocean sprawl a cause of jellyfish blooms? *Frontiers in Ecology and the Environment, 11*(2), 91–97. The article discusses how coastal development, pollution, and ocean sprawl contribute to the spread of jellyfish blooms, linking human infrastructure with changes in jellyfish distribution and abundance.

Gershwin, L. (2013). *Stung! On jellyfish blooms and the future of the ocean* (456 p). University of Chicago Press. A book that discusses the growing phenomenon of jellyfish blooms and their potential impact on marine ecosystems, fisheries, and human health. It also addresses the broader ecological changes that favor jellyfish dominance.

Lynam, C. P., Gibbons, M. J., Axelsen, B. E., et al. (2006). Jellyfish overtake fish in a heavily fished ecosystem. *Current Biology, 16*(13), R492–R493. This study highlights a case in the Benguela ecosystem where overfishing has led to jellyfish becoming the dominant species, illustrating the impact of reduced fish populations on marine ecosystem balance.

Pauly, D., Graham, W., Libralato, S., Morissette, L., & Deng Palomares, M. L. (2009). Jellyfish in ecosystems, online databases, and ecosystem models. *Hydrobiologia, 616*(1), 67–85. This paper provides an overview of the role of jellyfish in marine ecosystems and presents models that predict the consequences of jellyfish proliferation on ecosystem dynamics, fish populations, and food webs.

Pitt, K. A., & Purcell, J. E. (2009). Jellyfish blooms: Causes, consequences, and recent advances in research. *Hydrobiologia, 616*(1), 1–2. A brief overview of recent research into jellyfish blooms, emphasizing the causes of jellyfish proliferation and the ecological and socio-economic consequences of these blooms in marine environments.

Purcell, J. E. (2012). Jellyfish and ctenophore blooms coincide with human proliferations and environmental perturbations. *Annual Review of Marine Science, 4*, 209–235. This review links jellyfish and ctenophore blooms with human activities such as overfishing, habitat degradation, and climate change, highlighting how anthropogenic pressures are driving the rise of jellyfish populations.

Purcell, J. E., Uye, S. I., & Lo, W. T. (2007). Anthropogenic causes of jellyfish blooms and their direct consequences for humans: A review. *Marine Ecology Progress Series, 350*, 153–174. This review examines how human activities such as overfishing, pollution, and climate change contribute to the increasing frequency of jellyfish blooms and discusses the direct impacts on fisheries, tourism, and human health.

Richardson, A. J., Bakun, A., Hays, G. C., & Gibbons, M. J. (2009). The jellyfish joyride: Causes, consequences and management responses to a more gelatinous future. *Trends in Ecology & Evolution, 24*(6), 312–322. The article explores the factors driving the rise of jellyfish populations, the ecological and socio-economic consequences of jellyfish dominance, and potential strategies to manage jellyfish blooms in the future.

9

The Consequences of Invasive Species

Invasive species pose a significant and growing threat to marine ecosystems, driven largely by human activity. While the ocean has always been a dynamic environment where species migrate and interact, the introduction of non-native species by human means disrupts the delicate balance of marine ecosystems. These invasions can lead to the displacement of native species, the alteration of food webs, and long-term ecological consequences. The mechanisms by which invasive species enter new ecosystems are diverse, with global shipping, aquaculture, biofouling, the aquarium trade, and the construction of canals among the most common vectors. Ships, for example, take in ballast water in one port and release it in another, carrying with them species that may establish themselves in new environments. Aquaculture practices often result in the escape of non-native species, while biofouling involves marine organisms hitchhiking on the hulls of ships, allowing them to spread far beyond their native ranges. These vectors of invasion are exacerbated by the globalization of trade and travel, and with the increase in human activity, the potential for species invasions has grown exponentially.

The Case of *Mnemiopsis leidyi*

The invasion of *Mnemiopsis leidyi* (Fig. 9.1), a comb jellyfish native to the western Atlantic, into the Black Sea in the 1980s provides a well-documented case of how an invasive species can cause profound disruptions. It was introduced via ballast water discharge from ships, a common pathway for marine invasive species to spread to new regions. Once in the Black Sea, *Mnemiopsis*

Fig. 9.1 *Mnemiopsis leidyi*. (Author Albert Calbet)

found an ideal environment to thrive. With no natural predators and abundant food, primarily in the form of zooplankton, the comb jelly's population exploded.

The rapid proliferation of *Mnemiopsis leidyi* had devastating consequences for the Black Sea ecosystem. One of the most severe impacts was on the region's zooplankton populations, which were consumed in massive quantities by the invasive comb jelly. Zooplankton are critical components of marine food webs, serving as the primary food source for many fish species, particularly for larval and juvenile stages. By depleting this vital resource, *Mnemiopsis* starved native fish populations, including commercially valuable species such as anchovies (*Engraulis encrasicolus*) and sprat (*Sprattus sprattus*).

Anchovy populations were particularly hard-hit, leading to a sharp decline in fish catches. By the late 1980s and early 1990s, the anchovy stock had plummeted, causing economic devastation for the region's fisheries, which were highly dependent on this species. The collapse of fish stocks also had cascading effects throughout the ecosystem, as many marine predators, including larger fish and seabirds, depend on small pelagic fish like anchovies as a food source.

The collapse of anchovy and sprat fisheries in the Black Sea led to severe economic losses for coastal communities that relied on fishing as a primary source of income. According to estimates, fish catches in the Black Sea declined by more than 80% during the peak of the *Mnemiopsis* invasion, causing widespread unemployment and economic hardship for those involved in the fishing industry. The total annual loss to fisheries was estimated to be around $250 million USD during this period. The economic impacts extended beyond the fishing industry. Fish processing plants, local markets, and tourism sectors, which were linked to healthy fish stocks, also suffered significant losses. The decline in fish populations affected food security in the region as well, reducing the availability of affordable protein sources for local populations.

The *Mnemiopsis leidyi* invasion disrupted the Black Sea food web at multiple levels. As *Mnemiopsis* voraciously consumed zooplankton, it left little food available for native fish larvae, reducing the recruitment success of fish populations. This created a feedback loop where fewer fish reached maturity, further decreasing the abundance of predatory fish that could have helped control the jellyfish population.

Additionally, *Mnemiopsis* was highly efficient at outcompeting other species, reducing biodiversity and homogenizing the ecosystem. This species not only affected fish populations but also had indirect effects on phytoplankton. With fewer zooplankton grazing on them, phytoplankton populations increased, leading to eutrophication and the formation of harmful algal blooms. These blooms further degraded water quality and contributed to the development of hypoxic zones (areas with low oxygen), making conditions even more challenging for fish and other marine life.

To combat the *Mnemiopsis leidyi* invasion, scientists turned to biological control. In 1997, *Beroe ovata*, a comb jelly native to the western Atlantic that preys on *Mnemiopsis*, was introduced into the Black Sea. The hope was that *Beroe* would help restore ecological balance by reducing *Mnemiopsis* populations. The introduction of *Beroe ovata* had a noticeable positive effect. By preying on *Mnemiopsis*, *Beroe* helped reduce the comb jelly's numbers, allowing zooplankton populations to recover and easing the pressure on fish stocks. Within a few years, the *Mnemiopsis* population was significantly reduced, and fish catches began to rebound. Anchovy stocks, in particular, showed signs of recovery, and the economic situation for fisheries improved.

However, the introduction of *Beroe ovata* raised concerns about the long-term consequences of introducing yet another non-native species into the Black Sea. While the short-term results were positive, the potential for unintended ecological impacts remained. Introducing a new species to control an invasive one can lead to unforeseen consequences, such as shifts in predator-prey dynamics, altered competition between native species, and changes to the structure of the food web.

The Lionfish Invasion: An Expanding Crisis

Another well-known example of invasive species disrupting marine ecosystems is the lionfish (*Pterois volitans*, Fig. 9.2) invasion in the Atlantic and Caribbean. Native to the Indo-Pacific, lionfish were likely introduced to these waters through the aquarium trade, with one popular theory suggesting that they were accidentally released into the Atlantic during Hurricane Andrew in 1992. Since then, lionfish populations have exploded due to a combination of biological factors and environmental changes. With no natural predators in their new environment and a voracious appetite for reef fish, lionfish have caused severe declines in native fish populations. Their invasion has been particularly damaging to coral reef ecosystems, where lionfish prey on small reef fish that play critical roles in maintaining the health and biodiversity of the reefs.

Lionfish are generalist predators, meaning they consume a wide variety of prey, often targeting juvenile and small reef fish, such as parrotfish and wrasses,

Fig. 9.2 A large lionfish. (Image credit: Andrew David, NOAA/NMFS/SEFSC Panama City; Lance Horn, UNCW/NURC – Phantom II ROV operator)

which are crucial for maintaining coral reef health. Parrotfish, for example, graze on algae that can otherwise smother corals if left unchecked. Without these fish, algae overgrowth can significantly degrade coral reef ecosystems, leading to a loss of biodiversity and the collapse of reef structures. Studies show that in some regions, lionfish have reduced the biomass of native reef fish by up to 65% within just two years of establishing a population. This decline in native fish can cause cascading effects throughout the ecosystem, reducing biodiversity and altering the structure of the food web.

Additionally, lionfish are highly prolific, with females capable of releasing up to 30,000 eggs every few days throughout the year. This high reproductive rate, combined with their ability to thrive in a range of depths and habitats, makes lionfish extremely difficult to control. Their presence has been recorded from shallow coral reefs to deep-sea habitats, expanding their ecological footprint and making eradication nearly impossible through traditional fishing or hunting methods.

The lionfish invasion is no longer confined to the Atlantic and Caribbean; in recent years, it has expanded into the Mediterranean Sea. This expansion is attributed to climate change and the warming of Mediterranean waters, which have made the region more hospitable to Indo-Pacific species. The first recorded sighting of lionfish in the Mediterranean occurred in the early 2010s off the coast of Cyprus, and since then, their presence has steadily increased in various parts of the eastern Mediterranean, including Greece, Turkey, and Lebanon.

The invasion of lionfish into the Mediterranean is concerning for several reasons. First, the Mediterranean is already under pressure from overfishing, pollution, and climate change, making its ecosystems particularly vulnerable to invasive species. The introduction of lionfish could further exacerbate these issues by disrupting local food webs and reducing the populations of native species. Second, the Mediterranean is home to several endemic species that are not found anywhere else in the world. The presence of lionfish could threaten these species, leading to localized extinctions and a loss of biodiversity.

Studies have shown that lionfish in the Mediterranean exhibit similar predatory behaviors to those in the Atlantic, preying on a wide range of native fish and invertebrates. Some estimates suggest that lionfish populations in the Mediterranean could increase exponentially in the coming years, following the same trajectory as in the Atlantic. Without effective control measures, this invasion could lead to significant ecological and economic consequences, particularly for fisheries and coastal tourism.

The invasion of lionfish into both the Atlantic and Mediterranean has highlighted the challenges of managing invasive species in marine environments, where traditional control methods such as fishing or hunting can be difficult to implement on a large enough scale to be effective. In the Atlantic, efforts to control lionfish populations through organized fishing derbies, spear fishing, and commercial harvests have had limited success. Although these efforts have helped reduce local lionfish numbers in some areas, they have not been sufficient to halt the overall spread of the species.

In the Mediterranean, the consequences of unchecked lionfish populations could be severe. Fisheries in the region are already struggling due to overfishing and declining fish stocks, and the presence of lionfish could further deplete commercially valuable species. This is particularly concerning for small-scale coastal fisheries, which are vital to the livelihoods of many Mediterranean communities. Furthermore, the decline of native species could have broader ecological impacts, including the degradation of seagrass beds and coral reefs, which provide critical ecosystem services such as carbon sequestration, habitat for marine life, and protection against coastal erosion.

In addition to the ecological impacts, the rise of lionfish populations could have economic consequences for tourism in the Mediterranean. Many coastal communities rely on tourism, particularly diving and snorkeling, as a key source of income. Lionfish stings, which are painful and can be dangerous, may deter tourists from visiting affected areas, leading to economic losses for these communities. In the Caribbean, for instance, an increase in lionfish stings has been reported in popular diving destinations, raising concerns about the safety of recreational water activities.

Managing the lionfish invasion in both the Atlantic and Mediterranean will require a multifaceted approach. One potential solution is to promote the commercial harvesting of lionfish for human consumption. Lionfish are safe to eat, and their white, flaky meat is considered a delicacy in some regions. Promoting lionfish as a sustainable seafood option could help reduce their populations while providing economic opportunities for local communities. In the Caribbean, several restaurants have added lionfish to their menus, and organized lionfish fishing derbies have encouraged local fishers to target the species. Expanding these efforts to the Mediterranean could help mitigate the invasion while supporting local economies.

In addition to commercial harvesting, marine protected areas (MPAs) could play a role in controlling lionfish populations. MPAs are designated areas where human activities such as fishing are restricted or prohibited, allowing ecosystems to recover and thrive. Although MPAs alone cannot stop the spread of lionfish, they can help protect native species by reducing other stressors such as overfishing and habitat degradation. By maintaining healthy ecosystems, MPAs may help buffer the impacts of invasive species like lionfish.

Technological solutions, such as the development of lionfish-specific traps or autonomous robots designed to capture lionfish, are also being explored as potential tools for controlling their populations. However, these technologies are still in the experimental stages and may not be scalable enough to address the problem on a global scale.

The European Green Crab (*Carcinus maenas*)

The European green crab is native to the coastal waters of Europe, ranging from Norway and Iceland down to northern Africa. However, it has been introduced into numerous other regions, including the east and west coasts of North America, South Africa, Australia, South America, and parts of Asia. Its initial invasion of the United States was recorded as early as the 1800s, when the species is believed to have been transported across the Atlantic in the ballast water of ships.

Since then, the green crab has continued to spread, largely due to human activities such as shipping and aquaculture. Its ability to survive in a wide range of temperatures and salinities, combined with its adaptability and high reproductive rates, has allowed the species to colonize many diverse environments, including estuaries, rocky shores, and salt marshes.

The green crab is a highly aggressive and opportunistic predator that feeds on a wide variety of marine organisms, including bivalves (such as clams, mussels, and oysters), small fish, and other invertebrates. In areas where it has become established, the green crab has been shown to severely impact native species populations and disrupt local ecosystems.

One of the most significant impacts of the green crab invasion has been on shellfish populations. In the northeastern United States, for example, the green crab has been linked to the decline of soft-shell clam (*Mya arenaria*) populations. This has serious consequences not only for the local ecosystems, where clams play a key role in filtering water and maintaining sediment stability, but also for the region's economy. Shellfish harvesting is a major industry in many coastal areas, and the decline of clams and other commercially important species due to green crab predation has led to significant economic losses.

In addition to preying on shellfish, green crabs have been found to damage eelgrass beds, which serve as important nursery habitats for a variety of fish and invertebrate species. By uprooting eelgrass while foraging, green crabs contribute to habitat degradation, further impacting local biodiversity and reducing the resilience of coastal ecosystems to environmental changes such as rising sea temperatures and acidification.

The invasion of green crabs has resulted in severe economic consequences, particularly for fisheries and aquaculture operations. In the United States, the green crab has been responsible for the decline of the soft-shell clam industry, with losses in Maine alone estimated at millions of dollars annually. In other regions, the crab has negatively affected oyster and mussel farming, as it preys on juvenile shellfish, reducing stock abundance and increasing costs for aquaculture operators.

Efforts to control green crab populations through trapping and other methods have had limited success, and the cost of these management efforts adds further economic burdens on coastal communities. In Canada, for instance, both federal and provincial governments have invested heavily in green crab management, but the species continues to expand its range, affecting local industries and ecosystems.

The green crab's arrival in the Pacific Northwest of the United States and Canada in the late 1980s and 1990s has been a particularly concerning development. First detected in San Francisco Bay, the green crab spread rapidly along the Pacific coast, reaching British Columbia, Washington, and Oregon. The cold waters of the Pacific Northwest, initially thought to be inhospitable to the species, have proven suitable for green crab survival and reproduction, further threatening local ecosystems and fisheries.

In the Pacific Northwest, the green crab poses a direct threat to native species like the Dungeness crab (*Metacarcinus magister*), an important commercial and ecological species in the region. The green crab competes with juvenile Dungeness crabs for food and habitat, potentially reducing recruitment and affecting the overall health of the Dungeness crab fishery.

Efforts to manage the green crab invasion have focused primarily on trapping and monitoring programs to reduce population numbers and prevent further spread. In the United States, several states have implemented green crab trapping programs, often in collaboration with local fishers and researchers. These efforts have been moderately successful in reducing green crab populations in localized areas, but the species continues to thrive in many regions.

There have also been calls for increasing the commercial harvest of green crabs as a means of population control, similar to efforts with lionfish in the Atlantic. While green crabs are not widely consumed in most regions where they have invaded, some chefs and researchers have experimented with green crab recipes in an effort to promote them as a sustainable seafood option.

In Canada, scientists have explored the possibility of using biological control agents, such as parasites or diseases that specifically target green crabs, but these methods come with significant ecological risks. There is always the danger that introducing a new control species could have unintended consequences on native wildlife.

The long-term ecological consequences of the green crab invasion are still unfolding. As green crab populations continue to grow and spread, there is the potential for further declines in native species, including commercially important shellfish and ecologically significant species like eelgrass. The loss of biodiversity, habitat degradation, and continued pressure on fisheries will likely intensify as the species becomes more established in new areas.

The green crab invasion highlights the challenges of managing invasive species in marine environments, where control efforts are often costly, time-consuming, and only partially effective. As the species continues to expand its range, the need for innovative solutions and more comprehensive management strategies becomes increasingly urgent. The European green crab serves as a cautionary example of how invasive species can profoundly alter marine ecosystems and economies, underscoring the importance of preventing further invasions through stricter regulations on ballast water discharge, aquaculture practices, and other pathways that facilitate the spread of marine invaders.

Caulerpa taxifolia Invasion

Not only animals can invade new ecosystems, but also plants are able to colonize new areas with devastating consequences. One very famous and infamous example is that of *Caulerpa taxifolia*. Originally from tropical regions, *Caulerpa taxifolia* is a species of algae that was first identified as an invasive threat when a strain from a public aquarium in Monaco was accidentally released into the Mediterranean Sea in the early 1980s. This strain, sometimes referred to as the "aquarium strain," was adapted to colder waters and proved to be extremely resilient. *Caulerpa taxifolia* can grow in a wide variety of environments, including rocky substrates, seagrass meadows, and sandy bottoms, and it can thrive in both warm and cooler waters, making it highly adaptable.

Once released into the Mediterranean, *Caulerpa taxifolia* spread quickly, covering large areas of the seafloor with dense mats that smothered native seagrass species like *Posidonia oceanica*. By 2000, the invasive algae had colonized over 13,000 hectares of seabed along the coasts of France, Italy, Spain, and Croatia, and it continues to spread today.

The spread of *Caulerpa taxifolia* has had significant ecological consequences for Mediterranean marine ecosystems. One of the most notable effects is the displacement of native seagrass beds, particularly *Posidonia oceanica*, which is a critical habitat for many marine species. *Posidonia* meadows provide important ecological services, including acting as nurseries for fish and invertebrates, stabilizing sediment, and sequestering carbon.

Caulerpa taxifolia outcompetes native species by forming dense, fast-growing mats that block sunlight and reduce oxygen levels in the water. As a result, native seagrasses and algae are unable to grow, leading to a loss of biodiversity. Many fish and invertebrates that depend on seagrass meadows for shelter, breeding, and food are forced to move elsewhere or face population declines.

In addition to displacing native species, *Caulerpa taxifolia* has been found to produce toxic compounds that deter herbivores from grazing on it. This means that the algae can grow unchecked by natural predators, further contributing to its invasive success. The loss of biodiversity in *Caulerpa*-invaded areas can destabilize the entire marine ecosystem, with ripple effects throughout the food web.

The invasion of *Caulerpa taxifolia* also poses significant economic challenges for Mediterranean coastal communities. The dense mats of algae can hinder recreational activities such as swimming and boating, which impacts tourism, an important industry in the region. Moreover, *Caulerpa* can interfere with fishing activities by covering fishing grounds and making it difficult for fishers to deploy nets and traps.

The loss of biodiversity, particularly the decline of fish populations that rely on seagrass meadows, can also reduce the productivity of local fisheries, leading to economic losses for coastal communities that depend on fishing for their livelihoods. In some areas, the invasion has necessitated costly control and management measures, including manual removal of the algae and restrictions on fishing and recreational activities.

There are other species of algae and plants producing marine invasions. For instance, *Caulerpa cylindracea* (formerly *Caulerpa racemose*) has also become an invasive species in the Mediterranean, with its spread following a similar pattern. This species, like *C. taxifolia*, is able to rapidly colonize new environments and outcompete native species. It forms extensive mats that reduce biodiversity by smothering native seagrass and other plant species. *Caulerpa cylindracea* is particularly problematic because it can grow in a wide range of environments, including rocky substrates and deep waters. It has been found in areas up to 70 m deep, which makes it difficult to manage and control. Its ability to thrive in a variety of habitats makes it a major threat to Mediterranean ecosystems.

Another invasive marine plant that has caused significant ecological disruption is *Sargassum muticum*, a large brown alga native to the western Pacific

Ocean. It was introduced to European waters in the early twentieth century, likely through ballast water or oyster farming. Since then, *Sargassum muticum* has spread along the coasts of Europe and North America, where it outcompetes native seaweeds and alters coastal ecosystems. *Sargassum muticum* grows rapidly and can form dense canopies that block sunlight, reducing the growth of other marine plants and algae. This can lead to the displacement of native species and the alteration of entire ecosystems. In some areas, the presence of *Sargassum* has caused a decline in biodiversity, affecting fish, invertebrates, and other organisms that depend on native seaweed for habitat and food.

The Role of Aquaculture and Artificial Structures in Species Invasions

Aquaculture facilities (Fig. 9.3), harbors, and other artificial structures have increasingly become significant vectors for the spread of invasive species, as they provide ideal conditions for non-native organisms to establish themselves. These environments offer abundant food, shelter, and limited

Fig. 9.3 Aquaculture facility in Malaysia. (Author Albert Calbet)

predation, allowing invasive species to thrive and spread, often with devastating consequences for local ecosystems. As aquaculture and coastal development expand globally, their role as entry points for invasive species becomes a critical concern in marine conservation efforts.

Aquaculture, the farming of fish, shellfish, and other aquatic organisms, has grown exponentially in recent decades. The global aquaculture industry produced over 114 million metric tons of seafood in 2018, accounting for 52% of global fish consumption, and is projected to continue growing to meet the demands of an increasing global population. While aquaculture provides important economic and food security benefits, it has also facilitated the spread of invasive species.

One of the most prominent examples of invasive species spread through aquaculture is the Pacific oyster (*Crassostrea gigas*). Native to the northwestern Pacific, *C. gigas* was intentionally introduced to various regions, including Europe, North America, and Australia, for commercial aquaculture due to its fast growth and high market value. However, it quickly became invasive, displacing native oyster species and altering local ecosystems.

In areas such as the Wadden Sea (Germany) and the coasts of France, *C. gigas* has spread beyond aquaculture farms, establishing large populations that dominate intertidal zones. It competes with native species, such as the European flat oyster (*Ostrea edulis*), for space and resources. The invasive oyster forms dense reefs that disrupt natural sediment processes, alter habitat structure, and reduce biodiversity by outcompeting native bivalves and other organisms. These reefs also affect local food webs, as they can change the distribution and abundance of species that rely on native oyster habitats.

In Denmark, *C. gigas* expanded significantly in the Limfjord, leading to a decline in the native European oyster populations, which had previously supported local fishing industries. The establishment of *C. gigas* populations has also shifted ecosystem services, altering water filtration, sediment stabilization, and nutrient cycling.

Aquaculture not only spreads invasive animal species but also facilitates the introduction of invasive seaweeds. Species such as *Undaria pinnatifida* (wakame) and *Caulerpa cylindracea* have spread through aquaculture operations. *Undaria*, native to East Asia, was introduced to Europe, Australia, and New Zealand, often as a hitchhiker on aquaculture equipment or as part of shellfish farming practices. Once established, it spreads rapidly, displacing native seaweeds and altering coastal habitats by forming dense monocultures that reduce biodiversity and habitat complexity. *Caulerpa cylindracea* thrives in aquaculture environments and spreads by attaching to equipment or structures, often outcompeting native seagrasses and disrupting the ecosystem by altering nutrient cycling and providing poor habitat for native marine species.

Coastal development, including the construction of piers, seawalls, harbors, and marinas, creates artificial habitats that are ideal for invasive species. These man-made environments lack many of the natural features, such as predators or competitors, which would otherwise limit the spread of invasive species. These structures also provide surfaces for attachment and shelter, enabling invasive species to thrive and reproduce.

A study conducted along the coast of California revealed that artificial structures like docks, pilings, and seawalls harbor a higher density of invasive species compared to nearby natural habitats. In some cases, invasive species found on these structures—such as *Watersipora subtorquata* (an invasive bryozoan) and *Bugula neritina* (an invasive colonial bryozoan)—can spread to surrounding ecosystems, further exacerbating their impact.

Biocontrol Measures: Benefits and Risks

Biological control methods, such as the introduction of predators, have been employed as potential solutions to manage invasive species in marine environments. While these strategies can effectively reduce the populations of harmful invaders, they often carry significant ecological risks and require careful consideration before implementation. The introduction of *Beroe ovata* into the Black Sea is one notable example, but it also serves as a reminder of the complexities and uncertainties surrounding biological control measures. There is always the possibility that *Beroe ovata* could become invasive itself or that its presence could disrupt local food webs in unforeseen ways. For example, if *Beroe ovata* were to deplete *Mnemiopsis* populations to very low levels, it could impact other species that rely on *Mnemiopsis* as a food source, leading to further ecological imbalances. Additionally, *Beroe ovata* could potentially expand its range to other nearby ecosystems, where its introduction may have unintended consequences.

Biological control has been used in other contexts with varying degrees of success, and each case highlights the potential risks associated with introducing new species to manage invasions. For instance, to control the green crab populations, the United States and Canada have considered introducing predators, such as the native striped bass (*Morone saxatilis*), which preys on green crabs. While native striped bass may offer a natural solution, there are concerns that increasing their populations to control green crabs could disrupt local ecosystems. Striped bass are opportunistic feeders and may prey on other valuable species, such as juvenile lobsters, potentially exacerbating problems for native fisheries. This underscores the complexity of biological control: while introducing or promoting the population of native predators might seem like a safer option, there can still be unintended consequences for other species.

A classic example of biological control gone wrong is the introduction of the cane toad (*Rhinella marina*) in Australia. Although this is a terrestrial example, it serves as a powerful lesson for marine ecosystems. Cane toads were introduced in the 1930s to control pests in sugarcane plantations, but instead of controlling the pest species, the toads became highly invasive. They rapidly spread across northern Australia, poisoning native predators and disrupting ecosystems.

Although this case occurred on land, it highlights the danger of introducing a non-native species to control another. The impact of the cane toad continues to devastate Australia's biodiversity, and its presence underscores the need for caution when considering biological control measures in any ecosystem, including marine environments.

Another example of biological control involves the invasive sea squirt (*Ciona intestinalis*), which has established itself in several regions, including the North Atlantic and the Mediterranean Sea. This invasive tunicate outcompetes native species for space on coastal reefs and aquaculture facilities. Researchers have explored the possibility of introducing native predators, such as the sea urchin (*Paracentrotus lividus*), to manage *Ciona* populations. While sea urchins can consume *Ciona*, their introduction or promotion as a biocontrol agent could potentially disrupt local herbivorous species and alter algal communities, leading to unforeseen ecological consequences.

Biological control measures, while sometimes successful, come with inherent risks that must be carefully assessed. Some of the key challenges include the following:

1. *Unintended Ecological Impacts*: Introducing a predator to control an invasive species can lead to unexpected ecological consequences. The predator may not exclusively target the invasive species but could also prey on native species, thereby disrupting local food webs. For example, if a predator introduced to control an invasive fish species also preys on native fish species, it could further threaten already vulnerable populations. In marine ecosystems, where species interactions are complex and often poorly understood, introducing a new predator could have ripple effects throughout the food web.

2. *The Risk of the Biocontrol Species Becoming Invasive*: In some cases, the species introduced to control an invasive population may itself become invasive. This is a particular concern when the introduced species has no natural predators in its new environment. The cane toad in Australia is a clear example of this risk. Introduced to control agricultural pests, it became a highly invasive species, spreading far beyond its intended range and causing widespread ecological damage.
3. *Limited Effectiveness and High Costs*: Biological control is not always effective, and it can be difficult to manage once the introduced species has been released. In some cases, the biocontrol agent may not sufficiently reduce the invasive population, leading to continued ecosystem degradation. Additionally, the introduction of predators to control invasive species can be costly, and the long-term monitoring and management required to assess their impact can strain resources.

For example, efforts to control lionfish in the Caribbean and Mediterranean through hunting and fishing have had limited success because of the species' rapid reproduction and wide dispersal. While some localized control has been achieved, the costs of managing lionfish populations across large areas are prohibitive, and the species continues to spread.

The Urgent Need for Prevention and Management

It is widely accepted that "better safe than sorry" is an essential principle when it comes to managing the threat of invasive species. Preventing invasions in the first place is far more effective and less costly than attempting to control or eradicate an established population of invasive species. In marine environments, where species introductions can rapidly spiral out of control, prevention should be a top priority. This requires a multifaceted approach, including global cooperation, stricter regulations, improved biosecurity measures, and robust monitoring systems that are designed to detect and respond to invasions before they have a chance to cause irreversible harm.

One of the most significant pathways for the introduction of invasive species is ballast water discharge from ships. Ships take in ballast water in one region to maintain stability and then discharge it in another location, inadvertently introducing non-native species to new environments. This is how the comb jelly *Mnemiopsis leidyi* was introduced into the Black Sea, leading to ecological collapse in the 1980s. To prevent further introductions through this route, the International Maritime Organization (IMO) introduced the Ballast Water Management Convention in 2004, which requires ships to manage their ballast water to minimize the spread of invasive species. The convention, which came into force in 2017, mandates the treatment of ballast water through filtration, ultraviolet light, or chemical disinfection before it is released into the sea. This regulatory approach has been essential in reducing the risk of invasive species being transported through international shipping. Countries that have fully implemented ballast water management systems have seen a significant reduction in new invasions. For instance, Australia's strict enforcement of ballast water regulations has helped curb the spread of invasive species along its coasts, demonstrating the effectiveness of prevention strategies at a national level.

Another important vector for the spread of invasive species is aquaculture. While aquaculture provides an important source of food and economic revenue, it also poses risks when non-native species are farmed in open-water systems, serve the Pacific oyster as example. To mitigate these risks, several countries have implemented stringent biosecurity measures for aquaculture. In New Zealand, for instance, all imported species for aquaculture must undergo rigorous biosecurity screening to ensure that they do not carry invasive pathogens or have the potential to establish invasive populations. Additionally, strict controls on the transfer of aquaculture equipment between regions have been implemented to prevent the spread of invasive seaweeds, invertebrates, and other organisms that could hitchhike on the equipment. Such proactive measures have been key to preventing the spread of harmful species through aquaculture operations.

The aquarium trade is another vector for the introduction of invasive species into marine ecosystems, such as the lionfish in the Caribbean and Atlantic regions. To prevent similar invasions, the European Union has implemented strict regulations on the import and trade of potentially invasive marine species. For example, in the United States, some states have banned the sale and possession of lionfish, and there are similar regulations in parts of Europe aimed at preventing high-risk species from being imported. Education campaigns targeting aquarium owners have also been successful in raising awareness about the risks of releasing non-native species into the wild, thereby reducing the chances of future introductions. These measures demonstrate that the aquarium trade can be regulated effectively to prevent the spread of invasive species, provided there is a commitment to enforcement and public education.

While regulations and biosecurity protocols are critical components of prevention, early detection and rapid response are equally essential in stopping invasive species before they become entrenched in ecosystems. Monitoring programs that use advanced technologies, such as environmental DNA (eDNA) analysis, can detect the presence of invasive species in the water before populations have a chance to grow. eDNA sampling involves collecting water samples and analyzing the genetic material within them to identify species that are present, even if they are not visible. This technique has proven to be highly effective in detecting invasive species at low population densities, allowing for quicker intervention. For instance, eDNA monitoring has been used successfully in the Great Lakes region of North America to detect the early presence of Asian carp, an invasive species that threatens local fisheries. Once detected, rapid response teams can be deployed to contain and eliminate the invasive species before it spreads. The success of these programs depends on continuous funding and international collaboration, as invasive species often cross borders and require coordinated efforts to manage.

Invasive species management must take a proactive approach, focusing on prevention and early intervention rather than costly, often ineffective, attempts at eradication once species are established. Global cooperation is vital, as invasive species do not respect national borders and can spread through shared waterways and oceans. Stronger international agreements, such as the IMO's Ballast Water Management Convention, as well as regional partnerships, are essential for ensuring that prevention efforts are standardized and enforced across the globe. Additionally, local and national governments must take responsibility for implementing and enforcing biosecurity regulations that address their specific vulnerabilities, whether related to shipping, aquaculture, or other industries.

By focusing on prevention—through improved regulations, biosecurity measures, monitoring programs, and public education—countries can effectively minimize the risk of future invasions. Although the challenges posed by invasive species are formidable, proactive measures can safeguard marine ecosystems and protect biodiversity for future generations. Preventing invasions not only preserves the ecological integrity of marine environments but also prevents the economic costs associated with managing invasive species once they become established.

Further Reading

Albins, M. A., & Hixon, M. A. (2008). Invasive Indo-Pacific lionfish Pterois volitans reduce recruitment of Atlantic coral-reef fishes. *Marine Ecology Progress Series, 367*, 233–238. This study examines the ecological impacts of lionfish on Atlantic coral reefs.

Dejean, T., Valentini, A., Duparc, A., et al. (2011). Persistence of environmental DNA in freshwater ecosystems. *PLOS ONE, 6*(8), e23398. This study discusses the use of environmental DNA (eDNA) as a tool for monitoring invasive species in aquatic environments.

IMO (2017). *Ballast water management – The control of harmful invasive species*. International Maritime Organization. This document outlines the IMO's Ballast Water Management Convention and its role in preventing the spread of invasive species through global shipping.

Kideys, A. E. (2002). Fall and rise of the Black Sea ecosystem. *Science, 297*(5586), 1482–1484. This paper describes the collapse of the Black Sea ecosystem due to the Mnemiopsis leidyi invasion and its impact on fish populations, particularly anchovies.

Messing, R. H., & Wright, M. G. (2006). Biological control of invasive species: solution or pollution? *Frontiers in Ecology and the Environment, 4*, 132–140. This paper discusses the risks and benefits of using biological control to manage invasive species in marine environments.

Molnar, J. L., Gamboa, R. L., Revenga, C., & Spalding, M. D. (2008). Assessing the global threat of invasive species to marine biodiversity. *Frontiers in Ecology and the Environment, 6*(9), 485–492. This study presents a global assessment of the threats invasive species pose to marine ecosystems.

Schofield, P. J. (2010). Update on geographic spread of invasive lionfish (Pterois volitans and P. miles) in the Western North Atlantic, Caribbean Sea, and Gulf of Mexico. *Aquatic Invasions, 5*(Supplement 1), S117–S122. This paper tracks the spread of lionfish in the Atlantic, Caribbean, and Gulf of Mexico.

Shiganova, T. A., & Bulgakova, Y. V. (2000). Effects of gelatinous plankton on Black Sea and Sea of Azov fish and their food resources. *ICES Journal of Marine Science, 57*(3), 641–648. This article provides a detailed account of the ecological impacts of Mnemiopsis leidyi in the Black Sea, including its effect on zooplankton and fish stocks.

Williams, S. L., & Smith, J. E. (2007). A global review of the distribution, taxonomy, and impacts of introduced seaweeds. *Annual Review of Ecology, Evolution, and Systematics, 38*, 327–359. This review provides an overview of invasive seaweeds and the challenges of managing these species in marine environments.

10

The Future of Fisheries

For centuries, the world's oceans were seen as boundless sources of food and wealth. Coastal communities built their livelihoods around fishing (Fig. 10.1), which became deeply entwined with both economic growth and cultural identity. However, as humanity's footprint on the ocean has grown, the future of fisheries, once thought limitless, now teeters on the brink of collapse. Overfishing, habitat degradation, climate change, and pollution are straining marine ecosystems, putting both biodiversity and global food security at risk. As the ocean changes, so too must our approach to fisheries if we hope to secure a sustainable future.

The Dual Threat of Overfishing and Climate Change

Overfishing remains one of the most immediate threats to marine life. As we previously indicated, roughly 90% of global fish stocks are being fished at their maximum sustainable level (fully exploited) or beyond (overexploited/depleted). Industrial fishing fleets have scoured the oceans for decades, driving many species—especially large predatory fish like tuna, sharks, and swordfish—toward collapse. These species are not only commercially valuable but also crucial in maintaining the balance of marine ecosystems. Their decline leads to trophic cascades, where the loss of top predators allows prey populations to explode, disrupting entire ecosystems. In many cases, this has

Fig. 10.1 Traditional fishing at Hong Kong. (Author Albert Calbet)

contributed to an increase in jellyfish populations, which thrive in ecosystems where predatory fish populations have been decimated.

The depletion of these top predators has forced industrial fisheries to engage in "fishing down the food web," targeting smaller, less commercially valuable species such as sardines, anchovies, and even krill. This approach may temporarily sustain fishing industries, but it undermines long-term sustainability by depleting foundational species that support entire marine food webs. The ecological effects of this shift can be profound, as these smaller fish serve as critical prey for seabirds, marine mammals, and larger fish. Economically, this shift threatens to destabilize fishing industries, as once-profitable species become increasingly scarce, forcing fleets to travel farther and fish deeper, driving up costs and pushing many small-scale fishers into poverty.

Compounding these pressures is the impact of climate change, which is causing rapid shifts in fish populations. As ocean temperatures rise, many species are migrating to cooler waters, leaving behind traditional fishing grounds. Species like cod and haddock are retreating from the warming North Sea, while tropical species are expanding into temperate regions. For fishers, this means having to adapt to new species in unfamiliar areas, often with significant economic and social costs. Coastal communities that have relied on specific fish stocks for generations are seeing their livelihoods threatened as their target species move beyond reach.

Ocean acidification, another consequence of climate change, poses an additional challenge. As the ocean absorbs more CO_2, it becomes more acidic, affecting organisms that rely on calcium carbonate to form their shells and skeletons, such as oysters, clams, and certain species of plankton. The decline of these organisms would have devastating effects on marine ecosystems and the fisheries that depend on them. This is particularly troubling for shellfish farmers, who are already experiencing difficulties maintaining healthy stock in acidifying waters.

The Socioeconomic Divide: Industrial vs. Small-Scale Fisheries

The ecological damage wrought by overfishing and climate change has stark socioeconomic consequences, and the impacts differ dramatically between industrial and small-scale fisheries. Industrial fisheries, which dominate global seafood production, target high-value species using advanced technologies. These fisheries generate substantial economic output, but they often do so at the expense of marine ecosystems. By depleting fish stocks at unsustainable rates, industrial fleets jeopardize their own future while also putting immense pressure on coastal ecosystems. Industrial fishing is also responsible for large quantities of bycatch—the unintentional capture of nontarget species—leading to further declines in biodiversity and disrupting ecosystems.

Small-scale fisheries, by contrast, provide food and livelihoods for millions of people, particularly in developing nations. These fisheries typically use less-intensive methods and have a lower environmental impact. However, small-scale fishers are increasingly finding it difficult to compete with large industrial fleets, especially as fish stocks dwindle. For many of these communities, fish represent the primary source of protein and economic stability. As overfishing and climate change reduce the availability of traditional fish stocks, small-scale fishers are being pushed into poverty, and food insecurity is rising in regions that rely heavily on marine resources.

The Future of Seafood: A Changing Diet

As global fish populations continue to decline, the future of seafood and human diets is likely to shift toward smaller, more resilient species. Fish species like cod, tuna (Fig. 10.2), and salmon, which have long dominated global seafood markets, are becoming increasingly scarce and expensive due to overfishing, habitat destruction, and climate change. The Food and Agriculture Organization (FAO) reported that around 34% of global fish stocks were overfished in 2017, compared to just 10% in 1974. As these pressures

Fig. 10.2 A large tuna is landed by a group of fishermen. Image Courtesy of UN Food and Agriculture Organization and NOAA| (Author Danilo Cedrone)

continue to mount, the seafood industry and consumers will need to adapt to changing ocean ecosystems.

One alternative to overexploited species is turning to smaller, fast-reproducing fish like sardines, anchovies, and mackerel. These species are often more resilient to environmental changes because they have shorter lifecycles and can recover more quickly from population declines. Sardines and anchovies, for example, are highly efficient at converting plankton into biomass, making them a vital part of marine food webs and an important source of omega-3 fatty acids for human consumption. Their resilience has positioned them as more sustainable options for the future.

However, even these smaller species are not immune to the effects of climate change and overfishing. The Mediterranean sardine (*Sardina pilchardus*), once a key species for regional diets and economies, has experienced a steep population decline in recent years. Between 2008 and 2017, the biomass of sardines in the Gulf of Lion, a major Mediterranean fishery, dropped by about 75%. This decline is partly attributed to overfishing, but environmental changes play a significant role as well. Sardines feed primarily on plankton, which is sensitive to changes in sea temperature and nutrient availability. Rising ocean temperatures, driven by climate change, alter plankton communities, reducing the quantity and quality of food available for sardines and other small pelagic fish. As a result, sardines are not only shrinking in population size but also in physical size, with smaller fish being caught over time, further reducing their market value and ecological contribution.

Another example is the Peruvian anchovy (*Engraulis ringens*), which constitutes one of the world's largest fisheries and is primarily harvested for fishmeal used in animal feed. While anchovies are considered a more sustainable option due to their fast reproduction rates, they are heavily impacted by oceanographic changes associated with El Niño events. During strong El Niño years, warm waters disrupt the upwelling of cold, nutrient-rich water off the coast of Peru, leading to decreased plankton production and subsequent declines in anchovy stocks. In 1997–1998, one of the most severe El Niño events in recorded history caused a significant reduction in anchovy populations, with landings dropping by 50%.

The vulnerability of even resilient species like sardines and anchovies highlights the broader challenges of ensuring sustainable seafood in a changing climate. Shifts in ocean temperatures, acidification, and habitat degradation are reshaping marine ecosystems, pushing species beyond their historical ranges and altering their reproductive patterns. Moreover, marine heatwaves have increased by over 50% over the past century, further stressing fish populations and disrupting fisheries around the world.

In response to these pressures, future seafood diets may increasingly rely on aquaculture, which now supplies over 50% of the world's seafood. Additionally, alternative sources of protein from the ocean, such as seaweed and microalgae, are gaining attention as potential sustainable food sources.

Is Aquaculture a Sustainable Solution?

Aquaculture has long been seen as a promising solution to overfishing, providing a sustainable source of seafood without depleting wild fish stocks. However, as with many human activities, the reality of aquaculture's environmental impact is far more complex. Aquaculture, or the farming of fish and other marine organisms, is often touted as a sustainable alternative to wild fishing. Proponents argue that by cultivating species like salmon, shrimp, and shellfish in controlled environments, we can reduce the pressure on overfished populations and provide a reliable source of seafood to meet the demands of a growing global population. The Food and Agriculture Organization (FAO) estimates that by 2030, over 60% of the fish consumed by humans will come from aquaculture, a significant increase driven by both declining wild fish stocks and rising global demand for seafood.

However, despite its promise, aquaculture presents numerous environmental challenges. One of the primary concerns is the destruction of coastal habitats such as mangroves, seagrass beds, and coral reefs. In many parts of the world, these ecosystems are being cleared to make way for fish farms, a process that mirrors the deforestation seen on land. Mangroves, in particular, are crucial breeding and nursery grounds for a variety of marine species, and their removal can severely disrupt local biodiversity.

Another major issue associated with aquaculture is pollution. Fish farms generate significant amounts of waste, including uneaten feed, fish excrement, and chemicals such as antibiotics and antifoulants used to maintain the health of farmed species. This waste can accumulate in surrounding waters, leading to the creation of "dead zones" where oxygen levels are too low to support life. These zones often become breeding grounds for harmful algal blooms, which further reduce water quality and harm marine species. The Baltic Sea, for instance, has suffered from widespread eutrophication and hypoxia due to nutrient runoff, including waste from aquaculture facilities. In the Gulf of Mexico, nutrient runoff from agriculture and aquaculture has contributed to one of the largest dead zones on the planet, covering an area over 20,000 square kilometers.

In addition to pollution, aquaculture poses a risk to wild fish populations through the spread of diseases and parasites. Farmed fish, often raised in dense, confined conditions, are more susceptible to diseases like sea lice, which can easily spread to wild fish when they come into contact with farmed species. This is particularly concerning for migratory fish such as salmon, which pass by fish farms during their life cycles. The introduction of non-native species in aquaculture facilities can also lead to competition with local species, further destabilizing marine ecosystems.

A less visible but equally significant issue is the use of wild-caught fish to feed farmed species. Many aquaculture operations focus on raising predatory fish such as salmon, tuna, and shrimp, which require high-protein diets that often include fishmeal and fish oil made from smaller wild-caught fish like anchovies and sardines. These forage fish play a critical role in the ocean's food web, providing sustenance for seabirds, marine mammals, and larger fish. Overharvesting these species to feed farmed fish exacerbates the depletion of wild stocks and threatens the balance of marine ecosystems.

According to a 2018 report by the FAO, about 20% of the global fish catch is used to produce fishmeal and fish oil, much of which goes into aquaculture feed. This creates a paradox in which aquaculture, intended to alleviate pressure on wild fish populations, contributes to the depletion of those very populations through its reliance on wild-caught feed.

Despite these challenges, there is hope for a more sustainable future for aquaculture. Researchers and industry leaders are developing innovative solutions to address the environmental impacts of fish farming. One promising approach is integrated multi-trophic aquaculture (IMTA), which mimics natural ecosystems by farming different species together. In this system, waste from one species becomes a resource for another. For example, fish waste can be used to fertilize seaweed or feed shellfish, reducing pollution and creating a more balanced, self-sustaining system.

Another key area of innovation is the development of alternative feeds for farmed fish. Companies are exploring plant-based proteins, insect meal, and even lab-grown fishmeal as substitutes for wild-caught fish. These alternatives have the potential to significantly reduce the industry's reliance on forage fish and create a more sustainable food supply chain.

Recirculating aquaculture systems (RAS) are also gaining attention as a way to reduce the environmental footprint of fish farming. These land-based systems recycle water within the facility, minimizing waste discharge into the surrounding environment. RAS technology also allows for better control of water quality and reduces the risk of disease transmission between farmed and wild fish populations.

Climate change presents both challenges and opportunities for the future of aquaculture. Rising sea temperatures, ocean acidification, and extreme weather events can disrupt aquaculture operations, affecting fish growth, survival, and the overall productivity of farms. However, aquaculture can also play a role in mitigating climate change by providing a more sustainable source of protein than traditional livestock farming, which generates higher greenhouse gas emissions.

Scientists are also exploring the potential of seaweed and bivalve farming as climate-resilient forms of aquaculture. Seaweed absorbs carbon dioxide and nutrients from the water, helping to mitigate ocean acidification and eutrophication. Bivalves (Fig. 11.2), such as mussels and oysters, filter water as they feed, improving water quality.

As we look to the future, it is essential to recognize that aquaculture is not a panacea for the challenges facing our oceans. It must be part of a broader strategy that includes restoring wild fish populations, protecting critical habitats, and reducing pollution from all sources. By acknowledging the full scope of its impacts and embracing a holistic approach to marine conservation, we can ensure that aquaculture serves as a tool for sustainability rather than a contributor to ecological decline.

Socioeconomic and Ecological Consequences of Future Fisheries

The transition to a future of smaller, more resilient fish species and expanded aquaculture will have profound socioeconomic and ecological consequences. As high-value species become scarcer, the fishing industry will experience significant shifts. Industrial fisheries, which have long focused on large predatory fish, may need to adapt to harvesting smaller species. This will affect the global seafood market, potentially leading to higher prices for premium fish species and a shift toward more processed, lower-cost seafood products.

For small-scale fishers, the decline of traditional species and the rise of opportunistic species like jellyfish present a major challenge. Many of these fishers rely on specific species for their livelihoods, and the loss of these stocks could force them to turn to alternative, less profitable species. In some cases, small-scale fishers may be forced out of the industry entirely, leading to economic hardship and the loss of cultural traditions tied to fishing.

Ecologically, the shift toward smaller fish and jellyfish-dominated ecosystems will alter marine food webs. The decline of larger predatory fish reduces the ocean's ability to regulate prey populations, leading to imbalances that can cascade through entire ecosystems. The rise of jellyfish, in particular, creates "trophic dead ends" where energy is trapped within jellyfish populations, reducing the overall productivity of marine ecosystems and limiting the availability of food for other species.

Preventing Fisheries Collapse: A Path Forward

Preventing fisheries collapse is possible through a combination of strategic actions aimed at restoring balance to marine ecosystems and ensuring the long-term sustainability of seafood supplies (Fig. 10.3). One key approach is sustainable fisheries management, where governments must set and enforce science-based catch limits, protect critical habitats, and reduce bycatch to allow fish populations to recover. Marine protected areas are vital tools in this effort, providing refuges where ecosystems can regenerate and fish stocks can rebuild. Consumers also play an important role in promoting sustainable fisheries by diversifying their seafood choices and supporting certification programs like the Marine Stewardship Council. Combatting illegal, unreported, and unregulated fishing is another critical step in protecting fish stocks and maintaining global fisheries sustainability. International cooperation, satellite

Fig. 10.3 Fish market, Hong Kong. (Author Albert Calbet)

tracking, and blockchain technology can improve transparency and traceability in the seafood supply chain, making it harder for illegal fishing to thrive. Sustainable aquaculture is also a promising solution for meeting future seafood demand without further degrading marine ecosystems. Developing environmentally friendly aquaculture practices, including the use of alternative feeds and integrated farming systems, can help ensure that seafood production does not come at the expense of marine health. Additionally, the restoration of marine ecosystems, such as coral reefs, mangroves, and seagrass beds, is crucial for supporting healthy fish populations and enhancing the resilience of fisheries to environmental changes. These ecosystems serve as nurseries for many fish species and act as natural buffers against climate-related impacts. The future of fisheries is at a critical juncture, and the ocean's capacity to provide food and economic security is finite. Without significant changes to how marine resources are managed, the risk of depleting this vital resource is high.

Further Reading

FAO. (2018). *The state of world fisheries and aquaculture 2018: Meeting the sustainable development goals. FAO*. The FAO highlights aquaculture's growing role in global seafood production and addresses environmental challenges.

FAO. (2020). *The state of world fisheries and aquaculture 2020: Sustainability in action*. FAO. This FAO report provides up-to-date statistics on global fish stocks, sustainable practices, and the increasing reliance on aquaculture.

Froehlich, H. E., Gentry, R. R., & Halpern, B. S. (2018). Global change in marine aquaculture production potential under climate change. *Nature Ecology & Evolution, 11*, 1745–1750. This paper explores the future potential of aquaculture production as it adapts to global climate change.

Pauly, D., & Zeller, D. (2016). Catch reconstructions reveal that global marine fisheries catches are higher than reported and declining. *Nature Communications, 7*, 10244. This paper discusses the discrepancy between reported and actual fish catches, highlighting the overexploitation of marine resources.

Poloczanska, E. S., et al. (2013). Global imprint of climate change on marine life. *Nature Climate Change, 3*(10), 919–925. This research reviews the global impacts of climate change on marine species distribution and abundance.

Worm, B., Hilborn, R., Baum, J. K., Branch, T. A., et al. (2009). Rebuilding global fisheries. *Science, 325*(5940), 578–585. This key paper discusses the potential to rebuild global fisheries through effective management and regulation.

11

The Overexploitation of Sharks

A particular striking case of diversity loss because of overfishing is that of sharks. Sharks have long held a fearsome reputation, often portrayed as ruthless predators with an insatiable hunger for human flesh. This negative perception, fueled largely by popular media, especially movies like *Jaws* and its sequels, has painted sharks as dangerous creatures to be feared. However, this image is a distortion of reality. In truth, shark attacks on humans are extremely rare, with an average of only 72 unprovoked bites and around ten fatalities annually worldwide, according to the International Shark Attack File. On the other hand, human activities are responsible for the death of over 100 million sharks each year, primarily due to targeted fishing and bycatch.

This widespread fear and misunderstanding of sharks have contributed to a lack of urgency in their conservation. Yet, sharks are far more valuable alive than dead. They play a vital role in maintaining healthy marine ecosystems, and their overexploitation threatens not only shark populations but the stability of the entire oceanic food web. The future of sharks, and, by extension, the health of the ocean, is at risk due to human activity.

The Overexploitation of Sharks

The primary threat to sharks comes from human exploitation, largely driven by the demand for shark fins, meat, and other products. Shark finning is one of the most destructive practices, in which sharks are captured, their fins removed, and their bodies discarded back into the ocean, often still alive. Shark fins are highly valued in certain cultures, particularly in East Asia, where

Fig. 11.1 Typical store to buy shark fins in China. (Author Albert Calbet)

shark fin soup is a traditional delicacy (Fig. 11.1). This demand has driven a global trade that has decimated shark populations across the world.

A recent study published in *Science* in 2024 reveals that humans kill around 80 million sharks annually, with approximately 25 million belonging to threatened species. This represents a significant update to a widely cited 2013 study, which estimated shark mortality to be between 63 million and 273 million sharks per year, with 100 million being the most commonly cited figure. This includes both sharks targeted for their fins and meat and those caught as bycatch in fishing gear intended for other species. This staggering number has led to drastic declines in shark populations, with some species experiencing reductions of over 90%.

For instance, the oceanic whitetip shark (*Carcharhinus longimanus*, Fig. 7.2), once abundant in the open ocean, has seen its population drop by 98% in the Gulf of Mexico and by more than 90% in the Pacific. Similarly, hammerhead sharks (Sphyrnidae), particularly the great hammerhead and scalloped hammerhead species, have seen population declines exceeding 80% in many areas. As a result, the International Union for Conservation of Nature (IUCN) has classified over 37% of all shark and ray species as threatened with extinction.

Sharks are particularly vulnerable to overexploitation due to their life history traits. Many shark species take years to reach sexual maturity, have long gestation periods, and give birth to only a few pups at a time. For example, the great white shark (*Carcharodon carcharias*) can take up to 15 years to mature, and females typically give birth to two to ten pups. This slow reproductive rate makes it difficult for shark populations to recover from heavy fishing pressure.

The extinction of sharks would represent not just a tragic loss of biodiversity but would also have profound consequences for the functioning of marine ecosystems. Many species are already critically endangered, and unless current fishing practices change, we could lose some of the most iconic and ecologically significant predators in the ocean.

The Ecological Role of Sharks

Sharks are apex predators (Fig. 11.2), sitting at the top of the marine food web. They play a crucial role in regulating the populations of their prey, helping to maintain the balance of marine ecosystems. By preying on weaker, sick, or slower individuals, sharks promote healthy fish populations and prevent

Fig. 11.2 Whitetip shark. Location Hawaii, Papahanaumokuakea. (Courtesy of NOAA)

any one species from becoming too dominant. This regulation is essential for the biodiversity and resilience of marine ecosystems.

When sharks are removed from an ecosystem, the effects ripple throughout the food web, causing trophic cascades. Without sharks to control the population of smaller predators, these species can proliferate and overconsume their prey. For instance, in areas where predatory shark populations have been heavily depleted, there has been a notable increase in the populations of rays and skates, which feed on shellfish. In the Chesapeake Bay, the decline of sharks has allowed cownose rays to thrive, leading to the decimation of local scallop populations. This imbalance threatens not only the biodiversity of the region but also the livelihoods of those who depend on these resources.

Moreover, sharks indirectly contribute to the health of ecosystems like coral reefs and seagrass meadows by controlling the populations of herbivores and smaller predators that would otherwise overgraze or overhunt their habitats. Without sharks, coral reefs and seagrass meadows can become degraded, leading to further biodiversity loss and weakening the ecosystem's ability to sequester carbon, regulate climate, and support marine life.

Socioeconomic Impacts of Shark Declines

The collapse of shark populations has significant economic consequences, particularly for coastal communities that rely on marine tourism. Shark diving has become a major tourist attraction in countries like the Bahamas, South Africa, and Australia, generating millions of dollars in revenue each year. For example, shark tourism in the Bahamas is estimated to bring in $114 million annually, far exceeding the value of sharks in the commercial fishing industry.

Additionally, the collapse of shark populations can lead to the decline of commercially important fish species. By controlling the populations of smaller predators and herbivores, sharks play an important role in maintaining the health of fisheries. Without sharks, the unchecked proliferation of these species can disrupt the food web and lead to the depletion of other fish stocks, further exacerbating the challenges faced by coastal communities that depend on fishing for their livelihoods.

The Future Without Sharks: Ecosystem Collapse

The overexploitation of sharks has dire implications not only for the species themselves but for the future health of marine ecosystems. If shark populations continue to decline at the current rate, the consequences for ocean biodiversity will be catastrophic. The absence of sharks will trigger widespread trophic cascades, leading to the overpopulation of smaller predators, the depletion of key prey species, and the degradation of habitats such as coral reefs, seagrass meadows, and kelp forests.

The loss of sharks would fundamentally alter the structure and functioning of marine ecosystems. Without apex predators to regulate populations, ecosystems would become less stable and more vulnerable to other stressors, such as climate change and ocean acidification. This loss of resilience would make it increasingly difficult for ecosystems to recover from disturbances, pushing them closer to collapse.

The future without sharks is one of diminished biodiversity and weakened ecosystems. The services that healthy marine ecosystems provide—such as climate regulation, carbon sequestration, and the support of fisheries—would be severely compromised, with far-reaching consequences for both the environment and human societies.

What Can Be Done?

Despite the grim outlook, there are still opportunities to reverse the decline of shark populations and restore balance to marine ecosystems. Many countries are taking steps to protect sharks, either by banning shark finning, establishing shark sanctuaries, or implementing stricter fishing regulations. In some regions, marine protected areas offer refuges where shark populations can recover, and sustainable fisheries management practices are being adopted to limit bycatch.

However, much more needs to be done to ensure the survival of sharks. International cooperation is essential to combat illegal, unreported, and unregulated (IUU) fishing, which remains a major threat to shark populations. Stronger enforcement of existing protections, better monitoring of fishing activities, and public awareness campaigns are all critical components of shark conservation efforts.

Consumers can also play a role in shark conservation by choosing sustainable seafood options and supporting certification programs like the Marine

Stewardship Council, which promote responsible fishing practices. Additionally, ecotourism can provide a sustainable alternative to shark fishing, generating economic benefits for coastal communities while ensuring that shark populations remain healthy.

Further Reading

Clarke, S. C., McAllister, M. K., Milner-Gulland, E. J., Kirkwood, G. P., Michielsens, C. G. J., Agnew, D. J., et al. (2006). Global estimates of shark catches using trade records from commercial markets. *Ecology Letters, 9*(10), 1115–1126. This paper uses trade records to estimate the global scale of shark exploitation, particularly for the shark fin trade.

Davidson, L. N., Krawchuk, M. A., & Dulvy, N. K. (2016). Why have global shark and ray landings declined: Improved management or overfishing? *Fish and Fisheries, 17*(2), 438–458. This study investigates whether declining shark and ray landings are due to better management or overexploitation.

Dulvy, N. K., Fowler, S. L., Musick, J. A., Cavanagh, R. D., Kyne, P. M., Harrison, L. R., et al. (2014). Extinction risk and conservation of the world's sharks and rays. *eLife, 3*, e00590. This study provides an in-depth analysis of the conservation status of sharks and rays, highlighting the high extinction risk faced by these species.

Ferretti, F., Worm, B., Britten, G. L., Heithaus, M. R., & Lotze, H. K. (2010). Patterns and ecosystem consequences of shark declines in the ocean. *Ecology Letters, 13*(8), 1055–1071. This paper examines the patterns of shark population declines and their broader ecological consequences.

Shiffman, D. S., & Hammerschlag, N. (2016). Shark conservation and management policy: A review and primer for non-specialists. *Animal Conservation, 19*(5), 401–412. This review article provides an accessible overview of shark conservation policies and management strategies for non-experts.

Stevens, J. D., Bonfil, R., Dulvy, N. K., & Walker, P. A. (2000). The effects of fishing on sharks, rays, and chimaeras (chondrichthyans), and the implications for marine ecosystems. *ICES Journal of Marine Science, 57*(3), 476–494. This study reviews the impacts of fishing on sharks and related species, emphasizing how overfishing disrupts marine ecosystems.

Worm, B., Davis, B., Kettemer, L., Ward-Paige, C. A., Chapman, D., Heithaus, M. R., et al. (2013). Global catches, exploitation rates, and rebuilding options for sharks. *Marine Policy, 40*, 194–204. This research provides a global estimate of shark catches and exploitation rates, revealing unsustainable fishing practices.

12

Whales: Guardians of Marine Ecosystems and the Global Battle for Their Protection

Whales, often described as the giants of the ocean, serve not only as awe-inspiring symbols of the marine world but also as key contributors to the health and functioning of marine ecosystems. Their role in nutrient cycling, carbon sequestration, and maintaining balance in the food web is indispensable. Yet, for centuries, these majestic creatures have been hunted, nearly driven to extinction by human greed. Despite international efforts to curtail whaling, several countries continue to defy global regulations, highlighting the complexities of the struggle to protect whales.

Whales' Ecological Role

Whales (Fig. 12.1) contribute in vital ways to marine ecosystems, affecting nutrient cycling, productivity, and biodiversity. Known as the "whale pump," whales help distribute nutrients throughout the water column. When they dive to feed in the depths and return to the surface, they release nutrient-rich waste. This nutrient cycling is crucial, especially for phytoplankton—the base of the marine food web—which depends on nutrients like iron, nitrogen, and phosphorus. Phytoplankton, in turn, plays a pivotal role in oxygen production and carbon sequestration. While a large fraction of this oxygen is consumed in the ocean, marine phytoplankton still support life both within the ocean and on land.

As apex predators, whales help regulate prey populations, preventing overgrazing of marine plant life and maintaining balance within ecosystems. The presence or absence of whales can lead to significant shifts in food web

Fig. 12.1 Whale in Greenlandic waters. (Author Albert Calbet)

dynamics, a phenomenon known as "trophic cascading." In ecosystems with healthy whale populations, the balance between species and nutrient flows is stable. However, when whale populations decline, this balance can be disrupted, leading to unpredictable changes in the ecosystem.

The physical presence of whales, particularly their massive size, also plays a role in shaping marine habitats. For instance, whale carcasses, known as "whale falls," provide deep-sea habitats with an influx of nutrients, supporting diverse communities of scavengers and specialized organisms that thrive in nutrient-poor environments.

The Threats Facing Whales Today

Despite their ecological importance, whales continue to face a range of anthropogenic threats due to the various uses of products derived from their bodies. Historically, whaling was a major industry, driven by the demand for products such as whale oil, which was used for lighting lamps and as a lubricant, and baleen, which was utilized in the production of corsets, umbrellas, and other everyday items. Additionally, spermaceti, found in the head of sperm whales, was prized for making candles and machine lubricants, while ambergris, a rare substance from sperm whale intestines, was highly valued in the perfume industry. Although many of these products have been replaced by synthetic alternatives today, whaling persists in some countries for cultural, economic, and subsistence purposes, with whale meat and blubber still being consumed and traded.

By the time international attention turned toward conservation, whale numbers were critically low. Species such as the blue whale, once abundant in all oceans, were reduced to fewer than 3000 individuals by the mid-twentieth century. While commercial whaling has largely been banned by the International Whaling Commission (IWC), several countries continue to hunt whales. Japan, Norway, and Iceland are the primary nations that challenge the IWC's moratorium. Japan, until 2019, continued whaling under the pretext of scientific research, though much of the whale meat found its way into commercial markets. In 2019, Japan withdrew from the IWC entirely and resumed commercial whaling in its territorial waters, arguing for cultural preservation and economic necessity. Norway, which formally objected to the 1982 IWC moratorium, continues to hunt minke whales and sets its own quotas, while Iceland resumed whaling in 2006, also citing cultural and economic justifications.

Whale hunting has drastically diminished in most countries, but these few nations remain steadfast in their support of whaling, citing tradition and sustainability as their primary arguments. For example, Norway hunted over 500 minke whales in 2022, despite global outcry. Iceland, on the other hand, hunts both minke and endangered fin whales, making its whaling operations particularly controversial. The combined whaling activities of these countries not only threaten the recovery of whale populations but also undermine international conservation efforts.

In addition to hunting, whales face mounting threats from entanglement in fishing gear, ship strikes, noise pollution, and climate change. Entanglement in fishing nets is a significant cause of death for many whale species, including the critically endangered North Atlantic right whale, whose population has dwindled to fewer than 350 individuals. Ship strikes also pose a grave risk, particularly in busy shipping lanes like those off the coast of the United States, where whale migration routes intersect with commercial shipping activities. Efforts to reduce ship speeds and alter shipping lanes have been implemented in some regions to minimize these impacts, but more comprehensive measures are needed.

Noise pollution from underwater drilling, military sonar, and shipping traffic disrupts whale communication and navigation. Whales rely on sound for long-distance communication, especially for mating and locating prey, but human-generated noise can interfere with these essential behaviors. Disrupted communication can lead to disorientation, increased stress, and even strandings, particularly among species like sperm whales and beaked whales.

Lastly, climate change is altering the distribution of prey species, shifting migration patterns, and reducing the availability of key habitats. As sea temperatures rise and polar ice melts, the availability of prey such as krill and plankton—a critical food source for baleen whales like blue and humpback whales—is diminishing. These changes are forcing whales to travel farther and expend more energy to find sufficient food, adding to the stress they face from other human activities.

International Efforts to Protect Whales

While the threats to whales are significant, international conservation efforts have been gaining momentum. The establishment of marine protected areas (MPAs), such as the Southern Ocean Whale Sanctuary, provides safe havens where whaling is prohibited and human activities are limited. The Southern Ocean Whale Sanctuary, covering over 50 million square kilometers, has been instrumental in protecting species like the humpback whale, which has seen population increases in recent decades, thanks to these protections. However, sanctuaries are only effective if properly enforced, and countries like Japan have historically violated these protections by hunting within sanctuary boundaries under the guise of scientific research (Fig. 12.2).

Nongovernmental organizations (NGOs) such as Sea Shepherd and Greenpeace have played a crucial role in both protecting whales and raising awareness of the threats they face. Sea Shepherd, in particular, has actively intervened in illegal whaling operations, deploying ships to disrupt whaling fleets and prevent the capture of whales. Although controversial, these actions have helped draw attention to the issue of whaling and pressure governments to enforce conservation regulations more rigorously.

In addition to direct intervention, public education campaigns have helped shift global attitudes toward whales and whaling. Polling data from countries that practice whaling, such as Japan and Iceland, indicate that public support for whaling is declining, especially among younger generations. In Japan, the consumption of whale meat has dropped sharply, from over 200,000 tons per year in the 1960s to less than 5000 tons today. This shift in consumption patterns reflects a growing recognition of the ecological importance of whales and the moral implications of their continued exploitation.

At the same time, global legal frameworks such as the Convention on International Trade in Endangered Species (CITES) have helped regulate the

Fig. 12.2 Tying up the flukes of a finback whale to the bow of a commercial whaling vessel. (Image credit: NOAA Central Library Historical Fisheries Collection; FWS)

trade in whale products, making it more difficult for whaling nations to profit from their catches. CITES listings for various whale species ensure that international trade in whale meat, oil, and other products is closely monitored and restricted.

The Future of Whales and Marine Ecosystems

The future of whales is inextricably linked to the health of the marine ecosystems they inhabit. As key players in nutrient cycling and food web regulation, the loss of whales would have cascading effects on marine biodiversity. Whale populations have shown resilience and even recovery in regions where protections have been enforced, but the road to complete recovery is long and fraught with challenges. The increasing threats of climate change, ocean acidification, and habitat degradation add further urgency to the need for global action.

Efforts to save whales from extinction cannot be limited to banning whaling alone. Comprehensive measures are required to address the broader impacts of human activities on the ocean. This includes reducing bycatch, limiting ship strikes, mitigating noise pollution, and curbing carbon emissions to combat climate change. Only through concerted international cooperation and a shared commitment to protecting whales can these majestic creatures continue to thrive and play their vital role in maintaining the balance of the marine environment.

Further Reading

Baker, C. S., & Clapham, P. J. (2004). Modelling the past and future of whales and whaling. *Trends in Ecology & Evolution, 19*(7), 365–371. This article discusses the historical impact of whaling on whale populations and offers models to predict future trends in whale recovery.

Clapham, P. J., & Baker, C. S. (2002). Modern whaling. In W. F. Perrin, B. Wursig, & J. G. M. Thewissen (Eds.), *Encyclopedia of marine mammals* (pp. 1328–1332). Academic Press. This chapter provides an overview of modern whaling, including its practices, history, and current status.

Pershing, A. J., Christensen, L. B., Record, N. R., Sherwood, G. D., & Stetson, P. B. (2010). The impact of whaling on the ocean carbon cycle: Why bigger was better. *PLoS One, 5*(8), e12444. This paper examines how large whale populations historically contributed to carbon sequestration in the ocean, with implications for the global carbon cycle.

Reynolds III, J. E., & Perrin, W. F. (Eds.) (2009). *The biology of marine mammals* (861 p). Smithsonian Institution Press. A comprehensive textbook on the biology of marine mammals, covering topics from behavior to conservation.

Roman, J., & McCarthy, J. J. (2010). The whale pump: Marine mammals enhance primary productivity in a coastal basin. *PLoS One, 5*(10), e13255. This study demonstrates how marine mammals, particularly whales, help stimulate primary productivity in marine ecosystems through nutrient cycling.

Simmonds, M. P., & Isaac, S. J. (2007). The impacts of climate change on marine mammals: Early signs of significant problems. *Oryx, 41*(1), 19–26. This paper reviews the early effects of climate change on marine mammals, noting shifts in habitat, food availability, and population health.

Tønnessen, J. N., & Johnsen, A. O. (1982). *The history of modern whaling* (818 p). C. Hurst. A comprehensive historical account of modern whaling, detailing the development of whaling practices and their impact on whale populations.

13

Deep-Sea Ecosystems and Exploitation

The deep sea, comprising oceanic depths below 200 m, represents the planet's largest habitat, covering over 65% of Earth's surface. This environment is typified by extreme physical conditions: temperatures near freezing, immense pressure that increases by 1 atmosphere for every 10 m of depth, and complete darkness, save for the light produced by bioluminescent organisms. Despite these challenges, the deep sea supports a remarkable range of life forms, many of which remain undiscovered. The deep ocean is also a crucial player in global processes like carbon sequestration and nutrient cycling, maintaining the balance of marine ecosystems.

Increasing industrial interest in this previously unreachable realm is driven by the discovery of vast deposits of valuable minerals. Polymetallic nodules, seafloor massive sulfides, and cobalt-rich crusts are among the resources that could be exploited for use in electronics, renewable energy technologies, and other industries. However, this surge of interest raises concerns about the long-term environmental impacts of deep-sea mining, given the fragility of these ecosystems and the slow pace at which they recover from disturbances.

The Unique Deep-Sea Communities Based on Chemical Energy

While most life on Earth depends on sunlight and photosynthesis, deep-sea organisms often rely on chemosynthesis—a process by which energy is derived from chemical reactions, particularly those involving sulfur and methane.

Fig. 13.1 White smoky vent fluid rises out of small sulfur chimneys at Northwest Eifuku volcano. Pacific Ring of Fire 2004 Expedition. (NOAA Office of Ocean Exploration; Dr. Bob Embley, NOAA PMEL, Chief Scientist)

This process powers some of the most unique and isolated ecosystems on the planet, particularly those found near hydrothermal vents and cold seeps.

Hydrothermal vents (Fig. 13.1) were first discovered in 1977 along the Galápagos rift, marking one of the most significant discoveries in marine biology. These vents occur at tectonic plate boundaries where seawater seeps into the Earth's crust, becomes superheated by magma, and is expelled, laden with minerals like hydrogen sulfide. Specialized bacteria convert these chemicals into energy through chemosynthesis, forming the basis of a complex food web that includes organisms like giant tube worms (*Riftia pachyptila*), which grow up to 2 m long, mussels, and eyeless shrimp (Fig. 13.2) that cluster around vent openings. Vent communities can be found at depths exceeding 2500 m and are entirely independent of solar energy.

Cold seeps, another fascinating deep-sea ecosystem, release methane and hydrogen sulfide from beneath the ocean floor. Organisms inhabiting cold seeps, such as mussels, clams, and tube worms, also rely on chemosynthetic bacteria. These ecosystems tend to be slower growing than hydrothermal vent communities but can persist for much longer, sometimes for thousands of

Fig. 13.2 A dense bed of hydrothermal mussels, galatheid crabs, and shrimp. Northwest Eifuku volcano near a seafloor hot spring called Champagne vent. Pacific Ring of Fire 2004 expedition. (NOAA office of ocean exploration; dr. Bob Embley, NOAA PMEL, chief scientist)

years. Both hydrothermal vents and cold seeps host species that are slow to reproduce and long-lived, making them especially vulnerable to disruption.

Seamounts, underwater mountains rising from the deep-sea floor, also harbor unique ecosystems. They often act as biodiversity hot spots, home to deep-sea corals, sponges, and a myriad of fish species. Due to the strong currents around seamounts, these ecosystems rely on suspended nutrients and organic matter for sustenance. Yet, their slow growth rates mean that any disturbance, whether from deep-sea trawling or mining, can have long-lasting impacts.

The Threat of Deep-Sea Mining

Deep-sea mining focuses on three main resource types: polymetallic nodules, which contain manganese, nickel, copper, and cobalt; polymetallic sulfides, found around hydrothermal vents and rich in copper, gold, and silver; and cobalt-rich crusts, located on seamounts and essential for producing batteries. The demand for these minerals is driven by the need for renewable energy technologies, such as batteries for electric vehicles and wind turbines, as well as smartphones and other electronics.

Polymetallic nodules, for instance, are found scattered across the abyssal plains, particularly in the Clarion-Clipperton zone in the Pacific Ocean. This area is currently under heavy exploration by various countries and corporations, with estimated reserves of up to 1.2 billion tons of polymetallic nodules. Mining these resources involves the use of giant machines to scoop up nodules from the seafloor, causing massive sediment plumes that can spread over vast distances. These plumes can smother deep-sea organisms, clog feeding systems, and disrupt the local ecology.

Hydrothermal vent mining, meanwhile, would target the mineral-rich chimneys formed by vent activity. However, these vent communities are incredibly fragile and often take decades or centuries to recover from disturbances. Furthermore, mining these areas could destroy habitats that host species yet to be discovered, many of which may hold untapped potential for biotechnology, such as enzymes for industrial processes or novel pharmaceutical compounds.

Waste Disposal in the Deep Ocean: A Hidden Crisis

The deep ocean has also historically been used as a dumping ground for industrial waste, including hazardous materials like radioactive waste, chemical pollutants, and even carbon dioxide. During the cold war, from the 1940s through the 1970s, the United States, the Soviet Union, and several European nations disposed of large amounts of radioactive waste in the deep ocean. Estimates suggest that more than 100,000 barrels of radioactive waste were dumped in various oceanic locations, primarily the North Atlantic and the Pacific, at depths exceeding 4000 m. Although deep-sea currents move slowly, potentially limiting the dispersal of this waste, there is still concern that barrel corrosion could lead to leaks, posing long-term risks to marine ecosystems.

While radioactive dumping was officially banned in 1993 under the London Convention, legacy waste remains a significant concern. The deep-sea environment is often viewed as isolated and vast, leading to the assumption that it can "absorb" human pollution without consequence. However, recent studies have shown that pollutants, including microplastics, heavy metals, and persistent organic pollutants, have permeated even the deepest trenches of the ocean. For example, plastic debris has been found in the Mariana Trench, nearly 11,000 m below sea level, indicating that no part of the ocean is immune to human impact.

There has also been a growing interest in sequestering carbon dioxide in the deep ocean as a means of mitigating climate change. Proposals have been made to inject liquid CO_2 into the deep sea, where high pressure and low temperatures would theoretically keep it in a stable state. However, this approach raises significant ethical and ecological questions, as the long-term effects on deep-sea ecosystems are unknown. The introduction of large quantities of CO_2 could lead to acidification of localized areas, disrupting the delicate balance of life in these habitats.

The Legal and Regulatory Landscape

The regulation of deep-sea exploitation falls primarily under the United Nations Convention on the Law of the Sea, which designates the deep seabed beyond national jurisdictions as the "common heritage of mankind." The International Seabed Authority, established under United Nations Convention on the Law of the Sea, is responsible for issuing exploration licenses and regulating deep-sea mining activities. To date, the International Seabed Authority has granted over 30 exploration contracts to various countries and private companies, covering hundreds of thousands of square kilometers of the deep seabed.

Despite these regulatory frameworks, there are concerns that the pace of industrial activity is outstripping the ability of regulatory bodies to enforce environmental protections. The International Seabed.

Authority is under pressure to finalize regulations for commercial mining, yet the scientific community warns that we still lack a complete understanding of deep-sea ecosystems and their resilience to disturbance. Current legal frameworks also struggle to address the impacts of waste disposal, particularly historical radioactive dumping and the potential future use of the deep sea as a carbon storage reservoir.

Environmental groups and some nations have called for a moratorium on deep-sea mining until more is understood about the ecological consequences. Others argue that deep-sea mining is essential to meet the growing demand for critical minerals and to transition away from fossil fuels. Balancing these competing priorities will be a major challenge in the coming decades, as humanity continues to push into the deepest, least understood parts of the ocean.

Further Reading

Cordes, E. E., Jones, D. O. B., Schlacher, T. A., Amon, D. J., Bernardino, A. F., Brooke, S., & Levin, L. A. (2016). Environmental impacts of the deep-water oil and gas industry: A review to guide management strategies. *Frontiers in Environmental Science, 4*, 58. This review focuses on the environmental impacts of deep-sea oil and gas activities, including waste disposal and the potential risks to marine ecosystems.

Jones, D. O. B., Amon, D. J., & Chapman, A. S. A. (2018). Mining deep-ocean mineral deposits: What are the ecological risks? *Elements, 14*(5), 325–330. This article addresses the ecological risks associated with deep-sea mining, including habitat destruction and biodiversity loss.

Levin, L. A., & Sibuet, M. (2012). Understanding continental margin biodiversity: A new imperative. *Annual Review of Marine Science, 4*, 79–112. This review explores the biodiversity of deep-sea ecosystems and emphasizes the importance of understanding these habitats before they are impacted by human activities.

Miller, K. A., Thompson, K. F., Johnston, P., & Santillo, D. (2018). An overview of seabed mining including the current state of development, environmental impacts, and knowledge gaps. *Frontiers in Marine Science, 4*, 418. This paper provides an overview of the legal and regulatory challenges surrounding seabed mining, highlighting gaps in environmental protection and the role of international bodies like the International Seabed Authority.

Van Dover, C. L., German, C. R., Speer, K. G., Parson, L. M., & Vrijenhoek, R. C. (2002). Evolution and biogeography of deep-sea vent and seep invertebrates. *Science, 295*(5558), 1253–1257. This paper provides an overview of the unique ecosystems that exist around hydrothermal vents and cold seeps, focusing on the biogeography and evolutionary history of vent and seep organisms.

Wedding, L. M., Friedlander, A. M., Kittinger, J. N., Watling, L., Gaines, S. D., Bennett, M., & Smith, C. R. (2013). From principles to practice: A spatial approach to systematic conservation planning in the deep sea. *Proceedings of the Royal Society B, 280*(1773), 20131684. This paper discusses the potential impacts of deep-sea mining on marine ecosystems and the need for spatial conservation planning to mitigate those impacts.

Woodall, L. C., Robinson, L. F., Rogers, A. D., Narayanaswamy, B. E., & Paterson, G. L. J. (2015). Deep-sea litter: A comparison of seamounts, banks, and a ridge in the Atlantic and Indian oceans reveals both environmental and anthropogenic factors impact accumulation and composition. *Frontiers in Marine Science, 2*(art. 3). This study examines the extent of plastic and other waste in deep-sea environments, illustrating how human pollution has reached even the most remote ocean regions.

14

The Role of Marine Protected Areas

Marine protected areas (MPAs) have emerged as one of the most vital tools in the conservation of marine ecosystems, playing a crucial role in safeguarding biodiversity, promoting sustainable resource use, and mitigating the impacts of climate change. Over the past century, the development of MPAs has evolved from isolated conservation initiatives to a global movement aimed at addressing the myriad challenges facing the world's oceans. This chapter explores the historical development, current state, and future potential of MPAs, while highlighting their ecological, socioeconomic, and governance-related dimensions.

Defining Marine Protected Areas and their Importance

Marine protected areas are geographically defined regions in the ocean where human activities are regulated, restricted, or entirely prohibited to protect and conserve marine ecosystems. MPAs range from multiuse zones, where sustainable activities such as tourism and regulated fishing are allowed, to strict no-take zones, where all extractive activities are banned. The purpose of these protected areas is to safeguard critical habitats, protect endangered species, and preserve the natural processes that sustain marine life.

The establishment of MPAs reflects a growing understanding of the interconnectedness of marine ecosystems and the importance of maintaining healthy oceans. Overfishing, habitat destruction, pollution, and climate change have severely impacted ocean health, prompting governments and conservation organizations to designate MPAs as a means of counteracting these pressures. In addition to their role in biodiversity conservation, MPAs provide numerous ecosystem services, such as food production, carbon sequestration, coastal protection, and support for tourism industries.

Historically, indigenous coastal communities practiced informal forms of marine conservation by designating certain areas as off-limits for fishing during specific seasons, allowing fish populations to recover. These traditional practices have informed modern conservation strategies and highlight the long-standing importance of marine protection in maintaining ecosystem balance. Today, MPAs are recognized as one of the most effective tools for addressing the overexploitation of marine resources and the degradation of marine habitats.

The global extent of MPAs has expanded significantly in recent decades, now covering approximately 8% of the world's oceans. However, this is still far short of the 30% target set by the Convention on Biological Diversity (CBD) for 2030. Expanding the reach of MPAs will be essential for protecting marine biodiversity, particularly in ecologically significant areas such as the high seas and deep-sea environments, which remain vulnerable to industrial exploitation.

The Ecological Benefits of Marine Protected Areas

MPAs are integral to preserving marine biodiversity and enhancing ecosystem resilience. Numerous studies have demonstrated that well-managed MPAs, particularly no-take zones, result in significant increases in biomass, species abundance, and biodiversity. For example, data from the Great Barrier Reef Marine Park in Australia show that fish biomass within no-take zones is approximately twice as high as in adjacent fished areas, with particularly notable recoveries in large predatory fish such as coral trout. These findings underscore the role of MPAs in providing safe havens where overfished populations can recover and replenish surrounding areas through the spillover effect.

One of the most compelling ecological benefits of MPAs is the recovery of top predators, which are often the first species to decline due to overfishing. In Cabo Pulmo National Park, Mexico, where fishing has been prohibited since 1995, fish biomass increased by over 460% between 1999 and 2009, making it one of the most successful examples of marine recovery globally. The resurgence of top predators in this area has helped restore the balance of the local food web, demonstrating the broader ecosystem benefits of MPAs.

14 The Role of Marine Protected Areas

Fig. 14.1 Coral reef. Australia. (Author Albert Calbet)

In addition to supporting species recovery, MPAs protect critical habitats such as coral reefs, mangroves, seagrass meadows, and kelp forests. Coral reefs (Fig. 14.1), in particular, benefit from the protection offered by MPAs, as demonstrated by the Florida Keys National Marine Sanctuary, where coral cover in protected zones increased by 12% over 15 years, while coral in unprotected areas continued to decline. Similarly, MPAs in the Seychelles have improved coral reef resilience following mass bleaching events caused by rising sea temperatures.

Seagrass meadows and mangrove forests, which provide essential ecosystem services such as carbon sequestration and nursery habitats for juvenile fish, also thrive within MPAs. In Spain's Cabrera Archipelago Maritime-Terrestrial National Park, seagrass meadows expanded by 10% over a decade after being protected from coastal development and pollution. Mangrove ecosystems within the Belize Barrier Reef Reserve System have shown higher fish biomass in protected areas compared to unprotected mangroves, highlighting the critical role of MPAs in preserving habitats that support biodiversity and buffer coastal communities from natural disasters.

Deep-sea ecosystems, which are slow to recover from disturbances, are also safeguarded by MPAs. The South Georgia and South Sandwich Islands MPA, established in 2012, has allowed for the recovery of deep-sea species such as Patagonian toothfish, which had been overexploited by industrial fisheries. Protecting deep-sea habitats is essential for maintaining biodiversity in regions that are particularly vulnerable to industrial activities such as bottom trawling and deep-sea mining.

Marine Protected Areas and Climate Change

MPAs play a critical role in mitigating the impacts of climate change by protecting ecosystems that naturally sequester carbon and providing refuges for species adapting to changing environmental conditions. Blue carbon ecosystems—such as mangroves, seagrasses, and salt marshes—are highly effective at capturing and storing carbon dioxide. Mangrove forests, for instance, sequester carbon at rates three to five times higher than tropical rainforests. Protecting these habitats within MPAs helps maintain the ocean's capacity to absorb carbon and mitigate the effects of climate change.

In Madagascar, the Velondriake locally managed marine area has successfully protected mangrove forests, which not only support local fisheries but also sequester significant amounts of carbon. In Belize, the Turneffe Atoll Marine Reserve protects large areas of seagrass and mangroves, which act as critical carbon sinks. Studies in the region have demonstrated that these ecosystems contribute to reducing atmospheric CO_2 levels, highlighting the importance of MPAs in global climate mitigation efforts.

Beyond carbon sequestration, MPAs act as natural buffers against climate-related disasters. Coastal habitats like coral reefs, mangroves (Fig. 14.2), and salt marshes protect shorelines from storm surges, flooding, and coastal erosion, all of which are exacerbated by climate change. During the 2004 Indian Ocean tsunami, coastal areas shielded by intact mangroves and coral reefs experienced significantly less damage than regions where these habitats had been degraded. The Sundarbans Reserve Forest, a vast mangrove ecosystem in India and Bangladesh, protects millions of people from storm surges and rising sea levels while also serving as a crucial carbon sink.

MPAs also provide refuges for marine species affected by rising sea temperatures and ocean acidification. As waters warm, many species are forced to migrate to cooler regions or deeper waters. MPAs that span different latitudes

Fig. 14.2 Mangrove area in NE Australia. (Author Albert Calbet)

or depths, such as the Monterey Bay National Marine Sanctuary in California, offer critical migration corridors for species seeking more favorable environmental conditions. These connected areas help maintain biodiversity by allowing species to move and adapt in response to climate change.

Furthermore, MPAs can indirectly help mitigate some of the local impacts of ocean acidification, which threatens calcifying organisms such as corals, shellfish, and certain types of plankton. By reducing human-induced stressors like overfishing, habitat destruction, and pollution, MPAs create healthier and more resilient ecosystems, which are better equipped to cope with the challenges of acidification. Additionally, MPAs can protect critical habitats, such as seagrass beds and mangroves, which have the ability to locally buffer pH levels, providing some relief to nearby calcifying organisms. For example, the Channel Islands National Marine Sanctuary includes efforts to monitor the effects of ocean acidification on commercially important species, such as abalone and sea urchins, to better understand how to mitigate its impact on these vulnerable populations. Although MPAs do not directly stop acidification, they play a crucial role in preserving biodiversity and enhancing the resilience of marine ecosystems under changing environmental conditions.

Challenges in Marine Protected Area Management

Despite their ecological and climate-related benefits, MPAs face numerous challenges that can undermine their effectiveness. One of the most significant challenges is the lack of enforcement and monitoring, particularly in large or remote MPAs. Many MPAs suffer from "paper park" syndrome, where protections exist only on paper but are not enforced in practice. Illegal fishing, poaching, and unregulated activities continue in many MPAs due to insufficient resources for enforcement. It seems, for instance, that only 9% of MPAs in the Mediterranean have adequate enforcement measures, leaving much of the region's marine biodiversity unprotected.

To address these enforcement challenges, innovative technologies such as satellite monitoring and drones are being deployed to monitor remote areas and track illegal activities in real time. Programs like Global Fishing Watch allow authorities to monitor vessel movements and detect illegal fishing within MPAs, improving the ability to protect these areas from exploitation. However, the use of such technologies requires financial resources and technical expertise, which are often lacking in countries with limited budgets for conservation efforts.

Another major challenge is securing sufficient funding for MPA management. Effective conservation requires substantial investment in enforcement, monitoring, habitat restoration, and community engagement. Many MPAs rely on inconsistent funding from national governments or international organizations, limiting their ability to achieve long-term conservation goals. In Southeast Asia, for instance, more than half of the region's MPAs suffer from chronic underfunding, which hampers their effectiveness.

Socioeconomic conflicts also arise when MPAs are established without considering the livelihoods of local communities. In many coastal areas, fishing is a primary source of income and food security. Restricting access to fishing grounds through the creation of MPAs can lead to resistance from local populations, particularly when alternative livelihoods are not provided. In Indonesia, the designation of MPAs has faced opposition from artisanal fishers who rely on nearshore fishing for subsistence.

To address these conflicts, community-based management approaches, such as locally managed marine areas, have been developed to integrate traditional ecological knowledge with modern conservation practices. In Fiji, locally managed marine areas have empowered local fishers to take an active role in managing their marine resources, leading to improvements in both fish stocks and community livelihoods. Additionally, compensation schemes or alternative livelihood programs, such as ecotourism or sustainable aquaculture, can help ease the economic burden on communities affected by MPAs.

The dynamic nature of marine ecosystems presents another challenge for MPA management. Marine species often move across large distances, and activities outside MPAs can affect the ecosystems within them. For example, migratory species such as tuna are vulnerable to overfishing when they leave protected waters, highlighting the need for regional or global cooperation to manage migratory species effectively.

Connectivity and Technological Advancements in Marine Protected Areas

The effectiveness of MPAs is greatly enhanced when they are part of a well-connected network that considers the natural movement patterns of marine species and the interconnectedness of marine ecosystems. Many marine species, such as sharks, turtles, and seabirds, travel across vast oceanic distances for feeding, breeding, and sheltering. Ensuring the protection of these species throughout their life cycles requires linking multiple MPAs across national and international waters.

Creating networks of MPAs, particularly in regions where species migrate between feeding grounds, spawning areas, and nurseries, is critical to ensuring their protection. For example, the Coral Triangle, which spans six countries in Southeast Asia, has developed a network of MPAs that protect biodiversity across a wide geographic area. This network ensures that species such as tuna, sea turtles, and coral reef fish are safeguarded as they move between different habitats.

Genetic connectivity between MPAs is also crucial for maintaining the resilience of marine populations. Larval dispersal through ocean currents allows species to maintain genetic diversity, which is vital for adapting to changing environmental conditions. Coral reef ecosystems, for example, benefit from connectivity, as healthy reefs within MPAs can provide larvae to help repopulate degraded reefs in other areas. This process has been observed in the Great Barrier Reef, where protected reefs have supported the recovery of nearby damaged reefs through larval dispersal.

Technological advancements are revolutionizing MPA management by enabling more effective monitoring, enforcement, and data collection. Satellite technology, for example, allows for real-time monitoring of fishing vessels, making it easier to detect illegal activities within MPAs. Drones and autonomous underwater vehicles (AUVs) are also being used to patrol MPAs, collect data on biodiversity, and monitor the impacts of human activities such as trawling or mining.

Artificial intelligence and machine learning are increasingly being integrated into MPA management, helping to analyze vast amounts of data and predict illegal activities. These tools are making it easier for conservation authorities to allocate enforcement resources effectively and respond quickly to emerging threats. For instance, AI models are being used to predict coral reef resilience to bleaching, helping managers prioritize protection efforts in the most vulnerable areas.

Citizen science initiatives are also playing a growing role in MPA monitoring, with mobile apps allowing divers, snorkelers, and fishers to report sightings of marine species, illegal activities, or changes in environmental conditions. Programs such as Reef Check engage citizen scientists to monitor coral reef health worldwide, providing valuable data that informs MPA management decisions.

Integrating Marine Protected Areas into Global Ocean Governance

The successful protection of marine biodiversity through MPAs requires their integration into broader frameworks of global ocean governance. Many marine species and ecosystems transcend national borders, making international cooperation essential for effective conservation. Migratory species such as whales, sharks, and sea turtles cross multiple jurisdictions, requiring coordinated efforts between nations to protect them throughout their migratory routes.

The governance of the ocean is complex, with many areas beyond the jurisdiction of individual nations. The high seas, which cover more than 60% of the ocean's surface, are particularly challenging to protect due to the lack of a legal framework for biodiversity conservation in international waters. While the United Nations Convention on the Law of the Sea (UNCLOS) provides the foundation for ocean governance, it does not address the creation of MPAs on the high seas.

To fill this gap, negotiations are underway for a new international treaty focused on the conservation and sustainable use of marine biodiversity in areas beyond national jurisdiction (BBNJ). This treaty, often referred to as the "high seas treaty," would provide a framework for creating MPAs in international waters and ensure that marine resources are managed sustainably. Achieving international consensus on this treaty is critical for expanding MPA coverage into the high seas, which are home to a wealth of biodiversity, including deep-sea corals, hydrothermal vent communities, and migratory species such as tuna.

Regional cooperation is also vital for integrating MPAs into broader ocean governance frameworks. Regional Fisheries Management Organizations, which manage fish stocks and marine resources in international waters, play a key role in supporting the creation of MPAs. The Commission for the Conservation of Antarctic Marine Living Resources, for example, has established a series of MPAs in the Southern Ocean, including the Ross Sea region marine protected area, which covers over 1.55 million square kilometers. This MPA protects species such as penguins, seals, and whales while also serving as a critical biodiversity refuge.

MPAs also contribute to achieving the UN Sustainable Development Goals, particularly Goal 14, which focuses on conserving and sustainably using the oceans, seas, and marine resources. MPAs are essential for meeting the target of protecting 30% of coastal and marine areas by 2030, as outlined in the Convention on Biological Diversity's "30x30" initiative.

The expansion of MPAs into the high seas is essential for protecting marine biodiversity on a global scale. Several high-seas MPAs have already been established through regional agreements, such as those created by the OSPAR Commission in the North-East Atlantic. These MPAs protect important habitats such as seamounts and cold-water coral reefs, which support a diverse range of species.

The successful integration of MPAs into global ocean governance will require robust international cooperation, technological advancements, and enforcement mechanisms to ensure compliance with conservation regulations. Satellite monitoring, real-time data collection, and global surveillance networks will play a key role in managing MPAs on the high seas, where enforcement has historically been challenging.

Further Reading

Daigle, R. M., D'Aloia, C. C., Côté, I. M., Curtis, J. M. R., Guichard, F., & Fortin, M.-J. (2017). A multiple-species framework for integrating movement processes across life stages into the design of Marine Protected Areas. *Biological Conservation, 216*, 93–100. This study outlines a framework for incorporating species movement processes across different life stages into MPA design, emphasizing the importance of connectivity to ensure species persistence.

Dudley, N. (Ed.). (2008). *Guidelines for applying protected area management categories*. IUCN. This book provides comprehensive guidelines for managing MPAs according to the IUCN classification system, which helps define the levels of protection and management effectiveness.

Edgar, G. J., Stuart-Smith, R. D., Willis, T. J., Kininmonth, S., Baker, S. C., Banks, S., Barrett, N. S., Becerro, M. A., Bernard, A. T. F., Berkhout, J., et al. (2014). Global conservation outcomes depend on marine protected areas with five key features. *Nature, 506*(7487), 216–220. This study identifies five key features that lead to higher conservation benefits in MPAs.

Gerber, L. R., Botsford, L. W., Hastings, A., Possingham, H. P., Gaines, S. D., Palumbi, S. R., & Andelman, S. (2003). Population models for marine reserve design: A retrospective and prospective synthesis. *Ecological Applications, 13*(S1), S47–S64. This paper discusses the application of population models to inform the design of MPAs, emphasizing the need for larger protected areas to accommodate species with varying life histories and connectivity requirements.

Hilário, A., Metaxas, A., Gaudron, S. M., Howell, K. L., Mercier, A., Mestre, N. C., Ross, R. E., Thurnherr, A. M., & Young, C. M. (2015). Estimating dispersal distance in the deep sea: challenges and applications to marine reserves. *Frontiers in Marine Science, 2*, 6. This article addresses the challenges of estimating dispersal distances in deep-sea organisms and the implications for MPA design.

Jones, D. O. B., Amon, D. J., & Chapman, A. S. A. (2018). Mining deep-ocean mineral deposits: What are the ecological risks? *Elements, 14*(5), 325–330. This article addresses the ecological risks associated with deep-sea mining, including habitat destruction and biodiversity loss.

Sala, E., Costello, C., Dougherty, D., Heal, G., Kelleher, K., Murray, J. H., Rosenberg, A. A., Sumaila, U. R., & Pauly, D. (2013). A general business model for marine reserves. *PLoS One, 8*(4), e58799. This article provides a comprehensive model showing how well-designed and managed MPAs can generate economic benefits through fisheries recovery and increased tourism.

15

Emerging Technologies

As the ocean faces growing threats from climate change, overfishing, pollution, and habitat destruction, technological advancements are emerging as essential tools to protect, restore, and better understand marine ecosystems. These technologies hold the potential to transform ocean management, helping humanity balance sustainable use with conservation. However, these advancements come with significant challenges, and their implementation must be carefully managed to avoid creating new problems while solving others.

Satellite Monitoring and In Situ Monitoring: A Global View of Ocean Health

Satellite monitoring (Fig. 15.1) and remote sensing technologies have revolutionized ocean science, offering real-time data on a range of environmental factors, such as sea surface temperatures, plankton blooms, and illegal fishing activities. These tools have been critical for understanding large-scale changes in the ocean, including shifts in currents, the spread of plastic pollution, and even coral bleaching events caused by warming waters. Satellite imagery allows for the continuous monitoring of vast ocean areas, providing researchers with insights into how ecosystems are responding to climate change.

One of the most significant applications of this technology is in the monitoring of marine protected areas, where enforcement is often difficult due to the sheer size of the protected zones. Satellites can detect illegal fishing and

Fig. 15.1 Central portion of Florida, Gulf of Mexico seen from Gemini 11. (NASA)

pollution, improving the management of these vital conservation areas. While satellite data provides a macroview, it is limited in resolution, especially when it comes to deep-sea monitoring. Integrating this information with other sources of monitoring is key.

Recent advances in in situ autonomous ocean monitoring are transforming our ability to study marine environments with unprecedented precision and on a much larger scale. Autonomous platforms such as Argo floats, underwater gliders, autonomous underwater vehicles (AUVs), and sail-drones are now capable of collecting continuous, real-time data from remote ocean regions. The Argo program, initiated in the early 2000s, has deployed more than 4000 active floats worldwide, which measure temperature, salinity, and ocean currents from the surface down to depths of 2000 m. These floats transmit data every 10 days, contributing to an unparalleled global ocean monitoring network. Recently, the introduction of biogeochemical Argo floats, which measure additional parameters like oxygen, pH, nitrate, and chlorophyll, has further enhanced our understanding of ocean health. With an expected increase to 1000 biogeochemical floats by 2025, researchers are gaining critical insights into the ocean's role in regulating the Earth's climate, particularly through processes like carbon sequestration and acidification.

In addition to floats, underwater gliders are now a cornerstone of ocean monitoring. These gliders can travel distances of up to 5000 km and dive as deep as 1000 m, providing months of continuous data collection on a single battery charge. They are equipped with sensors that measure a range of oceanographic parameters, including temperature, salinity, dissolved oxygen, and chlorophyll fluorescence. Gliders have become crucial for tracking oceanic processes such as marine heatwaves and ocean stratification. For instance, in the Southern Ocean Carbon and Climate Observations and Modeling project, gliders have been deployed to gather data on carbon cycling and stratification, particularly in polar regions where climate change impacts are most acute. These data contribute to improving climate models and predicting future changes in ocean circulation patterns. The use of autonomous vehicles in regions like the Southern Ocean is particularly important given the logistical challenges of manual sampling in such harsh and remote environments.

Autonomous surface vehicles (ASVs), such as sail-drones, are also playing an increasingly vital role in ocean observation. These ASVs can travel up to 16,000 km over periods of 12 months, collecting high-resolution data on ocean-atmosphere interactions, carbon flux, and fish stock levels. For example, during a 2019 Arctic mission, sail-drones provided valuable real-time data on ocean acidification, sea surface temperatures, and carbon exchange processes between the ocean and atmosphere. These data have been integrated into models to better understand the impacts of climate change in the Arctic, a region particularly sensitive to warming. Additionally, sail-drones have been deployed in regions like the Southern Ocean to monitor seasonal changes in carbon uptake, which accounts for up to 30% of the global ocean's net carbon sequestration. This highlights the importance of autonomous monitoring in understanding the ocean's critical role in mitigating climate change.

Another transformative advancement is the use of environmental DNA (eDNA) sensors in situ, which allow for the detection of marine species by analyzing genetic material present in seawater. Autonomous eDNA samplers deployed in regions like the Northeast Atlantic have enabled real-time monitoring of marine biodiversity, tracking the presence of species such as bluefin tuna, sharks, and various plankton species. These sensors have proven to be more sensitive than traditional methods; a 2021 study found that eDNA sampling detected twice the number of species compared to conventional net-based techniques. This method is also more efficient and less invasive, reducing the need for large-scale physical sampling, which can disturb marine ecosystems. eDNA technology is proving essential in biodiversity monitoring, invasive species detection, and tracking species migrations, offering a new tool for conservation efforts.

Meanwhile, plankton imaging sensors are advancing our understanding of plankton dynamics by enabling continuous, detailed monitoring of plankton communities in their natural habitats. These sensors utilize cameras and image recognition algorithms to classify different plankton species and estimate their abundance in real time. For example, the Continuous Plankton Recorder (CPR) program, which has been operating for over 90 years, has amassed more than 300,000 samples from oceans worldwide. Through this technology, researchers have documented shifts in plankton populations in response to climate change. In the North Atlantic, plankton species have shifted 200–400 km northward over the past few decades, a trend closely linked to rising sea temperatures. Such data is invaluable for understanding how marine ecosystems are responding to global warming and nutrient changes, providing a historical baseline for assessing future changes. All of these technologies are generating vast amounts of data, which are being integrated with machine learning algorithms to process and interpret the information in real time.

Big Data and Predictive Modelling: The Digital Revolution

The integration of artificial intelligence (AI) into marine science is allowing researchers to process vast datasets with unprecedented speed and precision. AI algorithms can analyze satellite imagery, AUV data, and even acoustic recordings to detect illegal fishing, monitor species movements, or identify ecosystem shifts. For instance, AI-driven image recognition tools can detect fishing vessels operating in restricted zones, providing real-time alerts to enforcement agencies. AI is also making strides in biodiversity monitoring, particularly in identifying and tracking marine species like whales, sharks, and sea turtles. By recognizing specific patterns in ocean soundscapes, AI can detect species' vocalizations and movements, improving the management of endangered species and helping reduce bycatch in fisheries. While AI has immense potential, it is reliant on high-quality data. Poor inputs can lead to faulty analyses, and there are concerns about the ethical use of AI, especially regarding surveillance in territorial waters. Additionally, AI systems require significant energy resources, which presents a paradox in climate change mitigation efforts.

Despite the promise of big data and AI-driven models, there are limitations. AI systems can sometimes act as "black boxes," meaning their internal decision-making processes are not always transparent. This poses a challenge when AI models are used to inform critical conservation or management decisions, as scientists and policymakers may struggle to understand the rationale behind the predictions. Ensuring that AI models are interpretable and that their predictions are grounded in accurate, real-world data is essential to avoiding potentially harmful decisions.

Balancing Innovation with Traditional Expertise

While technological innovations offer exciting new opportunities for ocean monitoring and conservation, they must be grounded in traditional scientific knowledge. Fieldwork, taxonomy, and experimental ecology remain indispensable in providing the ecological context needed to interpret molecular and AI-driven data. For instance, while eDNA can confirm the presence of a species in an area, field observations are necessary to determine whether that species is thriving, declining, or merely passing through. Similarly, experimental ecology provides critical data on how species interact with their environment, which is essential for refining and validating AI-driven predictions.

The future of ocean sciences rests on a careful balance between cutting-edge innovation and traditional expertise (Fig. 15.2). As big data, AI, and molecular tools continue to evolve, they offer powerful new ways to monitor and protect marine ecosystems. However, these advancements must be integrated with the deep knowledge gained from decades of hands-on research, ensuring that technology is used not as a substitute but as a complement to traditional scientific practices. The most effective solutions for safeguarding the future of the ocean will come from a holistic approach that leverages the strengths of both technology and field-based expertise. In doing so, ocean sciences can continue to advance while preserving the fundamental principles that have guided marine research for generations.

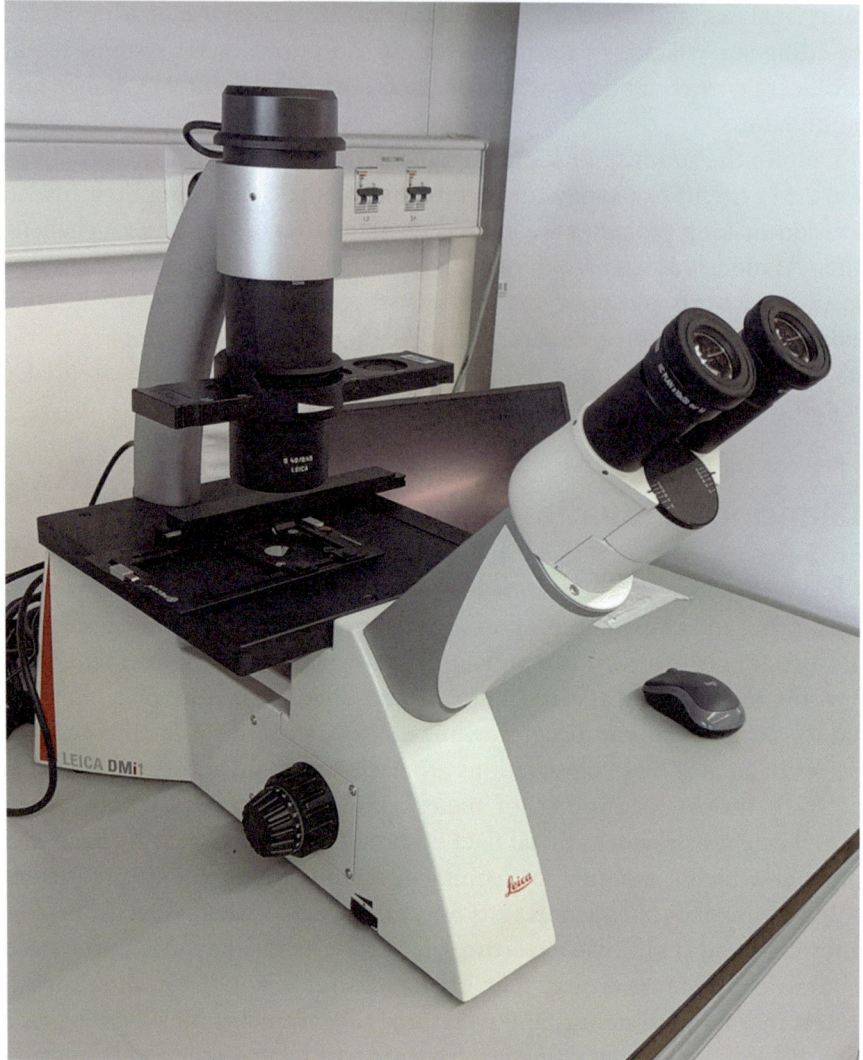

Fig. 15.2 Classic inverted microscope to observe plankton. (Author Albert Calbet)

Some Solutions Ahead: Ocean Cleanup and Carbon Sequestration

In the ongoing fight against plastic pollution, technologies like those developed by The Ocean Cleanup project are attempting to capture plastic waste in large, floating barriers to remove it from circulation. These systems are designed to trap plastic without harming marine life, but their effectiveness is still under evaluation. Cleanup projects are facing challenges, including difficulties in operating in rough seas and dealing with microplastics that cannot be easily captured.

Meanwhile, blue carbon ecosystems, such as mangroves, seagrasses, and salt marshes, are recognized for their significant role in carbon sequestration. These ecosystems are being restored using drones for replanting and GIS mapping for monitoring. While these efforts are crucial, they are costly and require long-term commitments. Additionally, blue carbon ecosystems are highly vulnerable to coastal development and climate change, which can negate their benefits if not protected.

As the pressure to reduce atmospheric CO_2 levels intensifies, various geoengineering proposals have emerged, many of which focus on enhancing the ocean's role as a carbon sink. Among the most controversial is iron fertilization, a process in which iron is added to nutrient-poor ocean regions to stimulate phytoplankton blooms. In theory, these blooms would absorb more CO_2, and some of this organic carbon would sink to the deep ocean, sequestering the carbon. However, this approach is fraught with risks, including the potential for harmful algal blooms and significant disruptions to marine ecosystems.

Other geoengineering concepts include artificial upwelling, which would involve pumping cold, nutrient-rich water from the deep ocean to the surface to boost phytoplankton growth, and alkalinity enhancement, which seeks to increase the ocean's ability to absorb CO_2 by reducing its acidity. These proposals are speculative, and their ecological and ethical implications are complex. Manipulating natural processes in such a drastic way could lead to unforeseen and possibly catastrophic consequences for marine ecosystems.

The growing emphasis on using the ocean to sequester carbon raises a critical concern: the ocean is becoming a container for all human trash—whether it be plastic waste, chemical pollutants, or excess CO_2. While the ocean plays a crucial role in regulating the global climate, relying on it to absorb ever-increasing amounts of human-generated CO_2 is not a sustainable solution. We must consider whether dumping our carbon waste into the ocean is a responsible approach or whether it merely delays addressing the root causes of climate change: overreliance on fossil fuels and unsustainable industrial practices.

One of the more direct methods of carbon removal involves carbon burial, capturing CO_2 from the atmosphere and injecting it into deep ocean sediments or storing it beneath the seafloor. While this could theoretically lock carbon away for millennia, it carries substantial risks. There is the potential for leakage, where stored CO_2 could escape back into the water column or atmosphere, undoing the benefits of sequestration. Additionally, injecting CO_2

into the deep ocean could further acidify local waters, damaging benthic ecosystems and marine species that are highly adapted to specific chemical environments.

The large-scale manipulation of ocean processes to enhance carbon sequestration—whether through artificial upwelling, fertilization, or carbon burial—poses serious questions. Are these methods merely technological distractions that allow continued fossil fuel use? The ethical considerations of using the ocean as a tool to fix our climate mistakes should not be taken lightly, especially given the uncertainties surrounding long-term impacts on marine ecosystems.

Technological advancements hold the potential to transform how we manage and protect the ocean. From remote monitoring to AI-driven conservation efforts, these innovations provide powerful tools to address the most pressing challenges facing marine ecosystems. However, they also present new risks and require thoughtful implementation. As we embrace these technologies, we must balance innovation with caution, ensuring that the solutions we deploy do not create additional harm or exacerbate the very problems they seek to solve.

In many cases, a holistic approach that integrates technological innovation with traditional conservation practices will be the most effective path forward. Whether it is using satellite data to enforce fishing regulations or deploying biotechnology to restore coral reefs, the future of ocean conservation will depend on how responsibly and effectively we leverage the tools at our disposal.

Further Reading

Chassignet, E. P., Pascual, A., Tintoré, J., & Verron, J. (2018). *New frontiers in operational oceanography GODAE OceanView* (811 p). This book provides in-depth knowledge of advancements in operational oceanography, focusing on the integration of satellite and in situ monitoring systems.

Evans, K., Lea, M. A., & Patterson, T. A. (2013). Recent advances in bio-logging science: Technologies and methods for understanding animal behavior and physiology and their environments. *Deep Sea Research Part II: Topical Studies in Oceanography, 88–89*, 1–6. This review highlights advancements in bio-logging technology, which allows the tracking of marine animals via attached sensors.

Jin, D., Hoagland, P., & Buesseler, K. O. (2020). The value of scientific research on the ocean's biological carbon pump. *Science of the Total Environment, 749*, 141357.

This article discusses the importance of oceanographic technologies in understanding the biological carbon pump and its role in climate mitigation.

Letessier, T. B., Juhel, J., Vigliola, L., & Meeuwig, J. J. (2015). Low-cost small action cameras in stereo generates accurate underwater measurements of fish. *Journal of Experimental Marine Biology and Ecology, 466*, 120–126. This study explores the use of low-cost cameras for underwater fish monitoring and habitat surveys.

Levin, L. A., & Le Bris, N. (2015). The deep ocean under climate change. *Science, 350*(6262), 766–768. This review discusses how new technologies, including AUVs and deep-sea sensors, help monitor the impacts of climate change on deep-sea ecosystems.

Lynch, P. D., Methot, R. D., & Link, J. S. (2018). *Implementing a next-generation stock assessment enterprise: An update to NOAA's stock assessment improvement plan.* NOAA Technical Memorandum NMFS-F/SPO-183. This document updates NOAA's framework for using emerging technologies like AI and machine learning in marine stock assessments.

Roemmich, D., & Gilson, J. (2009). The 2004–2008 Argo Program: Observing the global ocean with profiling floats. *Oceanography, 22*(2), 34–43. This paper provides an overview of the Argo program, which has revolutionized global ocean monitoring by deploying over 4000 autonomous floats to measure ocean temperatures, salinity, and currents.

16

Future Ocean Resilience and Adaptation Strategies

Despite the mounting challenges facing the world's oceans—climate change, overfishing, pollution, and habitat degradation—there is hope. Not everything is lost. The resilience of marine ecosystems, combined with innovative human-driven solutions, offers a way forward. Marine ecosystems have an innate capacity for recovery if given the chance, and human societies are increasingly recognizing the urgency of adopting sustainable practices. Scientific breakthroughs, community-based initiatives, and international cooperation provide pathways to a more sustainable future. While the window of opportunity to save our oceans is narrowing, there is still time to act decisively and preserve the vitality of the ocean for future generations.

Restoration of Coastal Habitats

Coastal ecosystems such as mangroves, seagrass meadows, and coral reefs provide critical services, including carbon sequestration, storm surge protection, and habitat for marine life. Yet, these ecosystems are highly vulnerable, with over 30% of the world's mangroves and 50% of coral reefs lost or severely degraded. Restoring these habitats is not only a strategy to enhance biodiversity but also a critical part of global efforts to mitigate and adapt to climate change. MPAs, discussed in Chap. 13, play a key role in safeguarding these ecosystems by reducing human-induced stressors and enhancing resilience to climate change. However, they are not the only helpful initiatives.

For instance, coral restoration projects have made strides in rebuilding degraded reefs. Coral gardening, where corals are propagated in nurseries and transplanted back onto damaged reefs, has shown promise in places like the Florida Keys and the Great Barrier Reef. Recent research has identified specific coral genotypes that are more resistant to heat stress and coral bleaching. For example, efforts in Australia are focusing on "assisted evolution," where corals are selectively bred or genetically enhanced to withstand higher temperatures. In these projects, corals are grown in controlled environments with elevated temperatures to encourage the development of heat tolerance before being transplanted to reefs.

Mangrove restoration, too, is proving effective at enhancing ecosystem resilience. Mangroves act as powerful carbon sinks, storing up to four times more carbon per hectare than tropical forests. Countries like Indonesia, home to some of the largest mangrove forests, are spearheading large-scale restoration efforts. In East Java, for example, over 100,000 hectares of mangroves have been restored, providing not only environmental benefits but also enhanced fisheries and improved coastal protection for local communities. The Global Mangrove Alliance aims to expand these efforts to protect and restore an additional 20% of mangroves by 2030.

Species Adaptation and Plasticity

Marine species are displaying remarkable plasticity—the ability to adapt to changing environmental conditions through a variety of mechanisms, including behavioral, physiological, and evolutionary changes. This capacity to adjust offers hope for certain species in a rapidly changing world. However, the extent and pace of these adaptations, alongside the unpredictability of climate change, mean that some species may still struggle to keep up.

One of the clearest examples of marine species' plasticity is the shift in their geographical ranges in response to ocean warming. As global sea temperatures rise, many species are migrating to cooler waters, often moving poleward or to greater depths where conditions are more favorable for their survival. For instance, a comprehensive 2020 study of North Atlantic fisheries revealed that over 50% of fish species have migrated either northward or deeper into the ocean. Atlantic cod (*Gadus morhua*), once abundant in the temperate waters of the North Atlantic, are now predominantly found at higher latitudes around Greenland and Iceland. This geographic redistribution reflects their need for cooler waters, as rising sea temperatures make their traditional habitats unsuitable for reproduction and survival.

16 Future Ocean Resilience and Adaptation Strategies

Similarly, other commercially important species, such as haddock (*Melanogrammus aeglefinus*) and herring (*Clupea harengus*), are experiencing significant shifts. The movement of these species is altering the composition of marine ecosystems and challenging fisheries management, as fish stocks move beyond traditional fishing grounds. This "climate migration" of marine life is not limited to fish. Species like lobsters, crabs, and squid are also migrating in search of cooler waters, forcing fishers to follow the new distribution of their target species. In the United States, for example, the American lobster has shifted northward from southern New England to the Gulf of Maine and Canadian waters due to warming seas.

While behavioral changes such as range shifts offer immediate survival benefits, evolutionary adaptations are also occurring in some species, providing a longer-term solution to environmental stressors. One notable example of this is seen in pteropods (Fig. 16.1), small marine snails that form a crucial part of

Fig. 16.1 Planktonic snail. *Limacina retroversa*. (Author R. Hopcroft)

the marine food web. Pteropods are particularly vulnerable to ocean acidification—a process where increasing CO_2 levels lower the pH of seawater, making it harder for these organisms to maintain their calcium carbonate shells. However, researchers have observed that some populations of pteropods are evolving thicker, more robust shells in response to acidification. This form of "evolutionary rescue" allows them to persist in increasingly acidic waters.

The rapid adaptation of pteropods is an encouraging example of how species can evolve to survive in a changing environment. However, this process is not guaranteed across all species, especially those with longer life cycles or lower reproductive rates. Furthermore, scientists caution that while some populations can evolve in response to environmental stress, not all species will be able to adapt quickly enough to the accelerated pace of ocean change. This raises concerns about the long-term survival of slow-reproducing species like corals, which are already facing severe threats from both warming waters and acidification.

In addition to range shifts and evolutionary adaptations, marine species are exhibiting physiological plasticity that allows them to cope with new environmental conditions. Fish, for example, have demonstrated remarkable flexibility in adjusting their thermal tolerance. Studies of coral reef fish, such as the damselfish (*Acanthochromis polyacanthus*), have shown that individuals exposed to elevated temperatures during their early developmental stages are able to adjust their metabolic rates, allowing them to tolerate higher temperatures later in life. This type of plasticity is crucial for survival in warming oceans, but its long-term sustainability is uncertain, especially as marine heatwaves become more frequent and severe.

Similarly, certain coral species have displayed some level of physiological plasticity by forming symbiotic relationships with more heat-resistant strains of zooxanthellae, the algae that live within coral tissues and provide them with nutrients through photosynthesis. These symbiotic shifts can help corals survive moderate increases in temperature, but they may not be sufficient to protect them from the more extreme warming predicted in the coming decades.

Behavioral changes also play a critical role in species adaptation to environmental change. As ocean ecosystems shift, some marine animals are altering their foraging and migration patterns to adapt to changes in prey availability. Seabirds, for instance, are exhibiting plasticity in their feeding behaviors. Puffins, which traditionally feed on fish like herring and sand eels, have been observed targeting different prey species such as squid and small pelagic fish as their usual food sources migrate to cooler waters.

Fig. 16.2 Whale in Greenland's waters. (Author Albert Calbet)

Marine mammals, particularly migratory species like whales (Fig. 16.2), are also adjusting their migration routes and feeding grounds in response to shifting prey distributions. For instance, humpback whales (*Megaptera novaeangliae*) are modifying their migratory paths and foraging areas to track the northward movement of krill and other prey species, which are responding to warming sea temperatures and changes in ocean currents.

As climate change accelerates and the adaptive capacity of many species becomes overwhelmed, conservationists are increasingly considering assisted migration—the deliberate relocation of species to more suitable habitats—as a potential solution. This strategy has been proposed for species that are unable to migrate or adapt quickly enough on their own, such as certain corals or coastal species facing habitat loss from sea-level rise.

However, assisted migration is not without controversy. While it could potentially save some species from extinction, there are risks associated with relocating species to new environments, including the potential for ecological imbalances or unintended consequences in the new habitat. Moreover, ethical concerns arise over the interventionist nature of such strategies, particularly when dealing with large-scale ecosystem disruptions.

Human Adaptation and Community-Based Solutions

Human societies, particularly coastal communities, are already experiencing the impacts of climate change on the oceans. Rising sea levels, changing fish populations, and increasing storm intensity are forcing coastal communities to adapt to new realities. In many cases, community-driven solutions are emerging as some of the most effective adaptation strategies.

Fisheries provide livelihoods for over 60 million people worldwide, and many coastal communities are reliant on fisheries for food security. As climate change shifts fish stocks and increases the frequency of extreme weather events, it is essential that fisheries management adapts to these new conditions. One approach to fostering climate-resilient fisheries is dynamic fisheries management, where catch limits, quotas, and fishing zones are adjusted in real time based on environmental data, such as sea surface temperatures and fish migration patterns. For instance, the Pacific Fishery Management Council in the United States has implemented dynamic closures for fisheries targeting Pacific sardines and hake, adjusting fishing pressure to match shifting population dynamics driven by temperature changes.

Nature-based solutions are gaining traction as effective strategies for both mitigating climate change and enhancing resilience. Coastal ecosystems like mangroves, salt marshes, and seagrasses are crucial for storing "blue carbon"— carbon sequestered in marine habitats. Blue carbon ecosystems can store between three and five times more carbon per hectare than terrestrial forests, making their protection and restoration a high priority for both climate mitigation and coastal protection.

Blue carbon projects are expanding rapidly. For example, in the Pacific Islands, a blue carbon initiative has been established to protect 20,000 hectares of seagrass meadows and mangrove forests, helping to offset regional carbon emissions while bolstering local fisheries and protecting coastal communities from storm surges. These projects often involve partnerships between governments, local communities, and the private sector, demonstrating the power of collaboration in tackling complex environmental challenges.

Global challenges require coordinated, international efforts. Policies that promote sustainability, equitable resource use, and environmental protection are essential for the future resilience of the oceans. The Paris Agreement, which aims to limit global warming to 1.5 °C, and the UN Sustainable Development Goals (SDGs), particularly Goal 14 ("Life Below Water"), provide frameworks for protecting ocean ecosystems.

One promising development is the High Seas Treaty, which was agreed upon in 2023 after years of negotiations. The treaty seeks to protect biodiversity in areas beyond national jurisdiction, covering nearly half of the Earth's surface. It establishes mechanisms for creating high seas marine protected areas, regulating deep-sea mining, and ensuring that marine resources are used sustainably. This treaty represents a major step forward in international ocean governance and demonstrates the growing recognition of the need to protect ocean ecosystems on a global scale.

Further Reading

Duarte, C. M. (2024). *Ocean secrets of the planet* (252 p). Springer Nature. This book provides basic facts for understanding the oceans, their properties, and their importance to mankind throughout history.

Duarte, C. M., Agustí, S., Barbier, E., Britten, G. L., Castilla, J. C., Gattuso, J.-P., Fulweiler, R. W., Hughes, T. P., Knowlton, N., Lovelock, C. E., et al. (2020). Rebuilding marine life. Nature, 580(7801), 39–51. Duarte and colleagues present a framework for rebuilding marine life by 2050, focusing on strategies such as fisheries management, MPAs, and pollution reduction.

Halpern, B. S., Frazier, M., Potapenko, J., Casey, K. S., Koenig, K., Longo, C., Lowndes, J. S. S., Rockwood, R. C., Selig, E. R., et al. (2015). Spatial and temporal changes in cumulative human impacts on the world's ocean. Nature Communications, 6, 7615. This research quantifies the increasing pressures on marine ecosystems, presenting data on how human impacts have intensified over time.

McCauley, D. J., Pinsky, M. L., Palumbi, S. R., Estes, J. A., Joyce, F. H., & Warner, R. R. (2015). Marine defaunation: Animal loss in the global ocean. Science, 347(6219), 1255641. This article discusses the large-scale loss of marine species due to human activities, emphasizing the need for better ocean stewardship.

Pauly, D., & Zeller, D. (2016). Catch reconstructions reveal that global marine fisheries catches are higher than reported and declining. *Nature Communications, 7*, 10244. This study highlights the underreporting of global fisheries catches, offering a comprehensive view of overfishing and its impacts on marine ecosystems.

Watson, J. E. M., Dudley, N., Segan, D. B., & Hockings, M. (2016). The performance and potential of protected areas. Nature, 515, 67–73. This paper assesses the effectiveness of Marine Protected Areas (MPAs) and the need for stricter enforcement of conservation measures.

Worm, B., Barbier, E. B., Beaumont, N., Duffy, J. E., Folke, C., Halpern, B. S., Jackson, J. B. C., Lotze, H. K., Micheli, F., Palumbi, S. R., & Sala, E. et al. (2006). Impacts of biodiversity loss on ocean ecosystem services. Science, 314(5800), 787–790. This landmark paper shows the consequences of declining biodiversity on marine ecosystems and services, linking species loss to ecosystem collapse.

17

Redefining the Human-Ocean Relationship

The ocean has always been a vital part of human life, from ancient civilizations that thrived along coasts to modern industries exploiting its depths for resources. But humanity's relationship with the ocean has reached a breaking point. Once seen as infinite and resilient, the ocean is now teetering on the brink of collapse. This chapter will examine the hard truths about how we have reshaped ocean ecosystems, manipulated resources, and polluted its waters. It is time to reconsider our relationship with the ocean and confront the consequences of our actions.

Humanity's Destructive Impact on the Ocean

Humanity's exploitation of the ocean has escalated from basic sustenance to destructive industrial activities. Fishing is perhaps the most glaring example of this transformation. While fishing used to be sustainable, modern industrial fleets now scrape the ocean floor, devastate fish populations, and destroy delicate habitats. Species such as cod, tuna, and sharks are on the verge of extinction. Coastal ecosystems that once nurtured life are overwhelmed by pollution and development. Our waste—including millions of tons of plastic and toxic chemicals—has transformed the ocean into a dumping ground for human excess.

The environmental degradation is driven by the false belief that the ocean is limitless. This illusion fuels a continued overexploitation of marine resources. The collapse of global fisheries, widespread destruction of coral reefs, and the emergence of dead zones are stark reminders of this damaging mindset. We are treating the ocean as a machine designed to serve our needs, ignoring the reality that it is a complex, interdependent ecosystem.

The Illusion of Progress and Infinite Growth

One of the most dangerous aspects of our relationship with the ocean is the illusion of progress—that we can continue growing our economies without fundamentally changing how we interact with the environment. Modern environmentalism often suggests that small actions, like recycling or switching to "sustainable" seafood, will solve the problems. However, these solutions are often superficial, masking the larger systemic issues that demand attention.

For instance, while many nations boast of creating marine protected areas, these zones are often inadequately enforced and underfunded. The global goal of protecting 30% of the ocean by 2030 sounds promising, but without serious enforcement, regulation, and political will, these efforts remain symbolic gestures. Plastic pollution is another area where superficial solutions dominate. Banning single-use plastics may make headlines, but the production of plastic continues to grow, and, at times, the alternatives are even worst. Microplastics are now embedded in marine food webs, with yet-unknown consequences for human health.

The idea of the "blue economy," which promotes sustainable ocean-based economic growth, reflects both the potential and danger of these superficial solutions. While renewable ocean energy and sustainable fisheries are essential, destructive industries like deep-sea mining and offshore drilling continue to expand. These practices threaten to devastate fragile ecosystems, disrupt nutrient cycles, and drive species to extinction—showing a deep contradiction between what we say we want and what we are actually doing.

Exploitation at the Heart of the Problem

Our political and economic systems are built on the premise of infinite growth and resource extraction, a mindset incompatible with sustainability. The industrialization of fishing is just one example: we have pushed many species to the brink of collapse, with little regard for the ecosystems that support them. The commodification of marine life, where fish are treated as products to be harvested, reflects our disconnection from nature.

The relentless pursuit of economic growth, however, extends far beyond fisheries. Deep-sea mining, for instance, threatens to devastate some of the least understood and most fragile ecosystems on the planet. Extracting minerals from the ocean floor could release vast amounts of stored carbon, disrupt critical nutrient cycles, and lead to the extinction of species that have yet to be studied. All of this is driven by the short-term pursuit of profits, with little understanding of the long-term consequences.

Hypocrisy in Ocean Management

There is a dangerous mix of complacency and hypocrisy in how we manage the ocean. Politicians often speak of sustainability while continuing to support policies that undermine it. Short-term economic interests take precedence over long-term environmental health, and many of the solutions proposed are half-hearted at best. For example, besides the already mentioned examples of marine protected areas and plastics, there are several other areas where hypocrisy in ocean management becomes evident. Overfishing remains a significant issue, with "sustainable" fisheries certifications often failing to uphold rigorous standards. Certified fisheries still engage in harmful practices (Fig. 17.1) such as bycatch and habitat destruction, creating a false sense of security for consumers. Similarly, while many governments promote green energy, they continue to expand offshore oil and gas drilling, which not only fuels climate change but also poses significant risks to marine ecosystems. The shipping industry, a major contributor to ocean pollution, often greenwashes minor improvements while continuing to operate under lax environmental regulations by using "flags of convenience." Deep-sea mining is promoted as a necessary step for the green energy transition, but this practice threatens fragile, largely unknown ecosystems, revealing a dangerous contradiction between conservation rhetoric and exploitative action. The tourism industry is another area rife with hypocrisy, as it markets pristine coral reefs and biodiversity while mass tourism and coastal development degrade these very ecosystems (Fig. 17.2). Similarly, while aquaculture is touted as a solution to overfishing, many fish farms contribute to pollution, disease, and overfishing of wild species used to feed farmed fish. Even blue carbon initiatives, such as restoring mangroves and seagrass to offset carbon emissions, often fall short, as companies use them to justify continued emissions elsewhere instead of addressing the root causes of climate change. These examples illustrate the deep contradictions and half-hearted efforts that characterize much of modern ocean management.

Fig. 17.1 Traditional sustainable fishing. (Author Albert Calbet)

Fig. 17.2 Dive tourism can place additional stress on reefs. Standing on or handling live corals can damage or even kill the coral polyps. (Image credit: David Burdick. NOAA Collection)

The Need for Political Will and Education

Addressing these challenges requires political leadership and systemic change. Incremental solutions will not be enough to halt the scale of environmental degradation we have caused. We must move away from a system that prioritizes short-term profits and growth over long-term sustainability. This involves holding corporations accountable for their environmental impact and ensuring they pay the true cost of their activities rather than externalizing them onto society and the planet.

To achieve meaningful change, we need leaders willing to acknowledge the full extent of the crisis and take bold, decisive action. This includes enacting policies that address the root causes of ocean degradation, such as overfishing, pollution, and climate change, rather than merely treating the symptoms. We must also strengthen international cooperation, ensuring that conservation efforts are equitable and that vulnerable coastal communities, particularly in developing nations, are not left behind.

Education will play a central role in reshaping the human-ocean relationship. Expanding ocean literacy programs can foster a deeper connection to the ocean and a sense of responsibility for its protection. By educating younger generations about the importance of the ocean, the threats it faces, and the steps needed to protect it, we can build a global movement committed to the long-term health of marine ecosystems.

At the individual level, people must take responsibility for their actions. Whether through reducing plastic consumption, supporting sustainable seafood, or advocating for stronger environmental protections, every action counts. Ultimately, the future of the ocean—and of humanity—depends on our willingness to embrace systemic change and adopt a philosophy of stewardship.

A New Relationship with the Ocean: Stewardship and Responsibility

Finally, we must also question our romanticized view of the ocean as an infinite frontier that can endlessly provide for human needs. This perception has led to the dangerous exploitation of marine ecosystems, mirroring the way we have exploited land-based environments. The reality is that the ocean, like forests, has limits to what it can provide, and pushing it beyond these thresholds can lead to the collapse of its ecosystems. For too long, the damage we inflict on the ocean has been hidden from sight—unlike deforestation, the destruction of coral reefs, overfishing, and pollution often occur out of view, making it easier for us to overlook the consequences of our actions. As a living, breathing ecosystem, the ocean is as crucial as any terrestrial environment in regulating climate, storing carbon, and supporting biodiversity. We must recognize that its beauty and mystery are not separate from its fragility. If we continue to view the ocean only as a resource, without acknowledging its limits, we risk destroying the very things we cherish most about it. Like forests, the ocean deserves the same urgency, respect, and protection.

Redefining our relationship with the ocean requires a profound shift in how we view it. We must move beyond seeing it as a resource to be exploited and begin treating it as a shared global responsibility. The shift is already underway, with initiatives like the UN Sustainable Development Goals (SDGs) and the growing emphasis on the "blue economy." However, for these efforts to succeed, they must be accompanied by real, systemic changes in how we manage and protect marine resources. A redefined human-ocean relationship must also address issues of equity and justice. Coastal communities are often the first to suffer from ocean degradation, yet they contribute the least to the problem. Conservation policies must prioritize the voices of these communities, ensuring that small-scale fishers have access to resources and that their rights are protected from industrial encroachment.

The future of the ocean is at a critical juncture. We have reached a point where our actions will determine whether marine ecosystems can continue to support life on Earth. This is not just an environmental issue but a matter of survival for humanity. We must abandon the destructive, exploitative practices of the past and embrace a new paradigm based on stewardship, sustainability, and equity. Only by confronting the uncomfortable truths about our relationship with the ocean can we hope to secure a future where both humanity and the ocean can thrive.

A Few Words of Hope

I did not want to end this book without offering some words of hope. While it is undeniable that we may never return to the ocean of 100 years ago—an ocean mostly untainted by the scale of human impact—this does not mean that all is lost. The ocean is resilient, and with concerted efforts, innovative science, and global cooperation, we can find ways to adapt, protect, and live in harmony with the ocean of the future. Through restoration, sustainable management, and embracing nature-based solutions, we have the tools to safeguard marine ecosystems for generations to come. The ocean may change, but it will continue to sustain life, enrich human cultures, and offer opportunities for recovery. Our challenge is not to recreate the past but to build a new relationship with the ocean—one rooted in respect, responsibility, and a shared commitment to ensuring that the future ocean remains a thriving and vital part of our planet.

Further Reading

Bernhardt, J. R., & Leslie, H. M. (2013). Resilience to climate change in coastal marine ecosystems. *Annual Review of Marine Science, 5*, 371–392. This paper reviews resilience in coastal ecosystems and how various factors contribute to their ability to recover from climate change impacts.

Doney, S. C., Ruckelshaus, M., Duffy, J. E., Barry, J. P., Chan, F., English, C. A., et al. (2012).Climate change impacts on marine ecosystems. *Annual Review of Marine Science, 4*(1), 11–37. A comprehensive review of how climate change is affecting marine ecosystems, focusing on resilience and adaptive responses.

García Molinos, J., Halpern, B. S., Schoeman, D. S., Brown, C. J., Kiessling, W., Moore, P. J., et al. (2016). Climate velocity and the future global redistribution of marine biodiversity. *Nature Climate Change, 6*(1), 83–88. The paper quantifies how fast marine species are moving toward cooler waters and the effects on biodiversity.

Munday, P. L., Warner, R. R., Monro, K., Pandolfi, J. M., & Marshall, D. J. (2013). Predicting evolutionary responses to climate change in the sea. *Ecology Letters, 16*(12), 1488–1500. This paper discusses evolutionary processes such as phenotypic plasticity and adaptation in response to climate change.

Pecl, G. T., Araujo, M. B., Bell, J. D., Blanchard, J., Bonebrake, T. C., Chen, I. C., et al. (2017).Biodiversity redistribution under climate change: Impacts on ecosystems and human well-being. *Science, 355*(6332), eaai9214. This review explores how the redistribution of species due to climate change affects ecosystems and human activities like fisheries.

Pörtner, H. O., & Farrell, A. P. (2008). Physiology and climate change. *Science, 322*(5902), 690–692. This paper focuses on physiological adaptations of marine organisms to climate change, including plasticity in thermal tolerance.

Sunday, J. M., Bates, A. E., & Dulvy, N. K. (2012). Thermal tolerance and the global redistribution of animals. *Nature Climate Change, 2*(9), 686–690. This study examines how shifts in thermal tolerance due to warming oceans are driving species redistribution.

MIX
Papier aus verantwortungsvollen Quellen
Paper from responsible sources
FSC® C105338

If you have any concerns about our products,
you can contact us on
ProductSafety@springernature.com

In case Publisher is established outside the EU,
the EU authorized representative is:
**Springer Nature Customer Service Center GmbH
Europaplatz 3, 69115 Heidelberg, Germany**

Printed by Libri Plureos GmbH
in Hamburg, Germany